The Asian Wild Man

The Asian Wild Man

**Yeti
Yeren
& Almasty**
Cultural Aspects & Evidence of Reality

Jean-Paul Debenat, PhD
Translated by Paul LeBlond, PhD

ISBN 978-0-88839-719-5
Copyright © 2014 Jean-Paul Debenat
First printing 2014

Library and Archives Canada Cataloguing in Publication

Debenat, Jean-Paul

[À la Poursuite du yéti. English]

The Asian wildman : yeti, yeren & almasty : cultural aspects & evidence of reality / Jean-Paul Debenat, PhD ; tranlated by Paul LeBlond, PhD.

Includes index.

Translation of: A la poursuite du yéti.

Issued in print and electronic formats.ISBN 978-0-88839-719-5 (pbk.).--ISBN 978-0-88839-720-1 (html)

1. Wild men--Asia, Central. 2. Sasquatch--Asia, Central. 3. Yeti.

I. LeBlond, P. H., translator II. Title. III. Title: À la poursuite du yéti.

English.

QL89.2.S2D4213 2014   001.944   C2013-908225-5

C2013-908226-3

All rights reserved. No part of this publication may be reproduced, stored in a retrieval system or transmitted, in any form or by any means, electronic, mechanical, photocopying, recording, or otherwise, without the prior written permission of Hancock House Publishers.

Printed in South Korea - PACOM
Editor: Theresa Laviolette
Production and cover design: Ingrid Luters

We acknowledge the financial support of the Government of Canada through the Canada Book Fund for our publishing activities.

Published simultaneously in Canada and the United States by
HANCOCK HOUSE PUBLISHERS LTD.
19313 Zero Avenue, Surrey, BC Canada V3S 9R9
1431 Harrison Avenue, Blaine, WA, USA 98230-5005
Tel: (604) 538-1114 Fax: (604) 538-2262

www.hancockhouse.com | sales@hancockhouse.com

Contents

Acknowledgements	7
Translator's Foreword	8
Preface	10
1. The Yeti in the 1950s	13
2. Nepalese Expeditions	20
3. The Book of Small People	29
4. The Yeti and Ethnomedicine	36
5. The In-between Times	43
6. Dasaï, Dasain or Dashain	47
7. In the Footsteps of a Myth	49
8. The Dumje Festival	53
9. Esau the Hirsute	57
10. Messner's Yeti	62
11. A Casual Meeting	66
12. Discoveries	69
13. Messner Perseveres	73
14. A Parenthesis: Links to the Pacific Northwest	76
15. Siberia	78
16. Kazakhstan	82
17. Mongolia	87
18. Diogenes in the Himalayas	91
19. Looking Back: Marie-Jeanne Koffmann	105
20. Mystery of the Braided Manes	111
21. Myth and Reality	114
22. Dr. Koffmann's Conclusions	119
23. The 1992 Almasty Expedition	124
24. Grover Krantz' Enquiry	130
25. The Wild Man in Modern China	134
26. Conclusion	144
Postscript	148
Appendix 1: From the Roof of the World to the Mesas of Arizona	149
Appendix 2: Angels and Demons	158
Bibliography	162
Index	167
About the Author	172
About the Translator	174

This book is dedicated to Maya
玛雅

Note on the term "almasty"
There are many terms in Russia for this unclassified hominoid. Every region in this vast country has its own name for the creature, and almasty (or almasti, almas) is simply one of the names. However, in the last 30 years or so, this word more than any others has been used for the creature in European and North American written material. I have therefore elected to use it as the common word for the Russian hominoid I discuss in this book.

Acknowledgments

Thanks to Peter Byrne (USA), Paul LeBlond (Canada), Christian Le Noël (France), Prof. Zhou Guoxing (China) and all those friends who helped me with my first book on the wild man (*Sasquatch/Bigfoot*, 2009, Hancock House).

I wish to thank the artists who have offered their illustrations: painter and writer Alika Lindbergh; and Philippe Coudray, illustrator and author, for his drawings of the greater and lesser yetis from his *Guide des animaux cachés* (Editions du Mont, 2009).

I am particularly grateful to my translator and friend, Paul LeBlond who I have known for many years. *The Asian Wild Man* is our second collaboration.

I am also most grateful to my faithful and patient secretary, my wife Marie-Agnès, and for the work and suggestions of my translator, Paul LeBlond.

Translator's Foreword

Are there really wild men still lurking in the mountains of Asia? Are the yeti of Tibet, the almasty of the Caucasus and the chuchunyas of Yakutia possibly manifestations of a relic population of primitive hominids, relegated into ever more inaccessible areas by the invading tide of their human competitors? Is their shyness and fear of humans the reason why they are seen so fleetingly and why their existence remains too doubtful to satisfy modern zoologists? Or are those vague and rare glimpses misinterpretations of bears and apes? Have those wild men perhaps by now vanished entirely, surviving only in the legends and myths of native populations?

Debenat takes us on a whirlwind tour of Asia, and its extension in the European Caucasus, in the footsteps of prominent wild man explorers—Peter Byrne, Marie-Jeanne Koffmann, Reinhold Messner, Zhou Guoxing—documenting their efforts to find answers to these questions. He complements historical and recent findings by searching the deep past through the window of mythology for traces of vanished races. Myths, religious rituals and folkloric events hold clues to the hidden meaning of the wild man, in Asia and elsewhere, as the author emphasized in his previous book on the sasquatch/bigfoot.

The possibility of survival of another branch of humanity, parallel to and estranged from *Homo sapiens*, continues to fascinate scientists and novelists. Were our mysterious Neanderthal cousins and competitors exterminated, as most believe; assimilated, as some recent evidence suggests; or merely pushed aside into the wilderness? In this book, Debenat presents a panorama of the state of knowledge of the wild man in Asia, which spans the continent in space and the centuries in time and which leaves the reader with a deeper appreciation of one of the most fascinating mysteries of modern anthropology.

Originally published in French, this book draws from a wide variety of sources; in adapting it to a North American readership, citations have been made to English language references wherever available. Cultural allusions obvious only to a Gallic audience have been adapted or explained where necessary.

Paul LeBlond
Galiano Island, BC

Preface

> He had barely begun with the traditional expression, "*Once upon a time…*" when a woman interrupted him.
> *Just what time are you talking about?*
> He answered, *That time when the animals spoke.*
> There were of course smiles and whispers. Thinking himself clever, a man said, *You mean at THAT time?*
> *Exactly*, answered the storyteller. *But please do not interrupt me again: I can't explain and tell the story at the same time. It's your choice.*
>
> —Jean-Charles Pichon, from *L'âne qui a vendu son maître*
> (The donkey who sold his master), 1996

A few years ago, my many travels, readings and encounters left me with an extensive documentation about an elusive creature: the slippery and evasive sasquatch, or bigfoot, the hairy giant of the American Pacific Northwest. I gathered my notes in a book wherein I also, on occasion, referred to the wild man of Asia and to *Gigantopithecus*, bigfoot's widely debated putative ancestor.

In spite of occasional difficulties arising from their governments' politics, investigators in North America, Russia and China have always exchanged information. Right from its creation by Bernard Heuvelmans in the 1950s, cryptozoology attracted the curious everywhere and even eminent scientists. The International Society of Cryptozoology (ISC), over which Heuvelmans presided, included among its honorary members Marjorie Courtenay-Latimer, Marie-Jeanne Koffmann, Théodore Monod, Sir Peter Scott and John R. Napier.

In 1938, Marjorie Courtenay-Latimer, curator of the East London Museum in South Africa, discovered a fish covered with bony plates rather than scales, a living fossil thought extinct for 65 million years: the coelacanth! Local fishermen knew of it as a food fish called *gombassa* or *mame*. This discovery, at first met with incredulity, caused a major sensation in zoological circles.

As to Dr. Marie-Jeanne Koffmann, the reader will learn about her and her works below in the chapters relating to the almasty, the wild man of the Caucasus.

Théodore Monod, was a specialist of desert ecosystems, an erudite and prolific author as well as a poet and philosopher. His works are a trove of information for scientists as well as a source of inspiration for all thinking people.

Sir Peter Scott, prominent ornithologist and nature painter, was one of the founders of the World Wildlife Fund and the artist who designed its famous panda logo.

Physician, primatologist and paleoanthropologist John Russell Napier was the Director of the Primate Biology Program at the Smithsonian Institution. He was interested in the wild man phenomenon and wrote about it in *Bigfoot: the Yeti and Sasquatch in Myth and Reality* (1973).

More recently, the renowned chimpanzee specialist Jane Goodall has also shown an interest in the wild man.

These few examples should suffice to show that people interested in cryptozoology are certainly neither ignorant nor feeble-minded. The lure of mystery also draws in a broad public to works describing creatures either still unknown or believed to have disappeared. The authors of such works regularly repeat the same information since they often draw from each other's books. The most honest of them of course quote their sources. As for myself, I have sought to go a little further or, in any case, to go back in time. I have strived to draw from the very sources of information by putting the emphasis on the fundamental works of the pillars of cryptozoology, be they field investigators or theoreticians, or perhaps both at the same time as were the zoologist Bernard Heuvelmans and the physician Marie-Jeanne Koffmann.

I also speak in this study of the work of Vera Frossard, a videographer whose documentary on the yeti has, in my view, not received enough attention. I will also delve at length into the research carried out by the alpinist Reinhold Messner, wrongly decried by some, and I will defend his excellent report on the yeti. I will also quote from the works of a Russian-born Englishwoman, Odette Tchernine, a poet and essayist to whose works Bernard Heuvelmans introduced me many years ago. Many quote from her pioneering work; too few acknowledge it.

I also turned my attention to China, where a wild man, the yeren, has often been reported. A meeting with Prof. Zhou Guoxing opened the door to that investigation and further correspondence with him allowed me to broaden my knowledge.

Finally, in Appendix 1, I report on links between the bigfoot *kachina* of the Hopi and Tibetan traditions. I thus wished to loop the loop with a brief foray back to North America.

My literary training, as well as my keen interest in traditional cultures, together with the friendly and insightful teachings of Bernard Heuvelmans and of the prominent writer and mythologist Jean-Charles Pichon, has led me to emphasize a particular perspective: the hidden meaning of the creature described as the wild man. In order to decipher the deeper symbolism of that creature, I have drawn upon mythology, the history of religions (notably shamanism), and even esoteric knowledge.

I have of course taken into account factual data (field observations, scientific descriptions...), but I have described as best as I can the domain where I feel most competent: interpreting the meaning hidden in tales, legends, rituals, masks and other cultural artifacts.

I invite the reader to discover, as I have, some unexpected aspects of the wild man in Asia.

Pont St. Martin
January 2011

1. The Yeti in the 1950s

A few years ago, Peter Byrne and Robert Short were reunited for a special ceremony. They were celebrating the 60th anniversary of the crash of a twin-engine Catalina seaplane in a lagoon of the Cocos Islands (also known as Keeling Island, 800 kilometers [497 mi] southwest of Java). The aircraft, part of Squadron 240 of the Royal Air Force based in Madras, carried 14 Canadian and British airmen when it fell into the lagoon on June 27, 1945. Nine people died. Peter Byrne, then in the RAF, was a member of the rescue team first on the scene. On June 27, 2005 he and Robert Short, the last remaining survivor of the crash, took part in a commemorative ceremony on West Island, one of the Cocos.

This rather solemn function is a vivid reminder of what an adventurous life Peter Byrne has led. In 1946, while he was still in uniform, he first heard about the "snowman" from Sherpas and Lepchas[1] during a leave in Darjeeling in the mountains of northern India. Demobilized in London in 1947, he set out to investigate the mysterious mountain primate. He discovered that except for some Chinese manuscripts dating back to 200 BC, there was hardly any mention of it until 1832. In that year, the British resident at the royal court of Nepal in Katmandu wrote that during an expedition his men had been scared by a *rakshas*,[2] a Sanskrit word still used in the 1970s. The tribes of northern Nepal were also said to use the term *shookp* or *sogpa* to denote the yeti. In Byrne's words:

> The Resident, Mr. B.H. Hodgson, a well known naturalist in his day, described the creature seen by his men as walking erect, tailless and covered with long dark hair.[3]

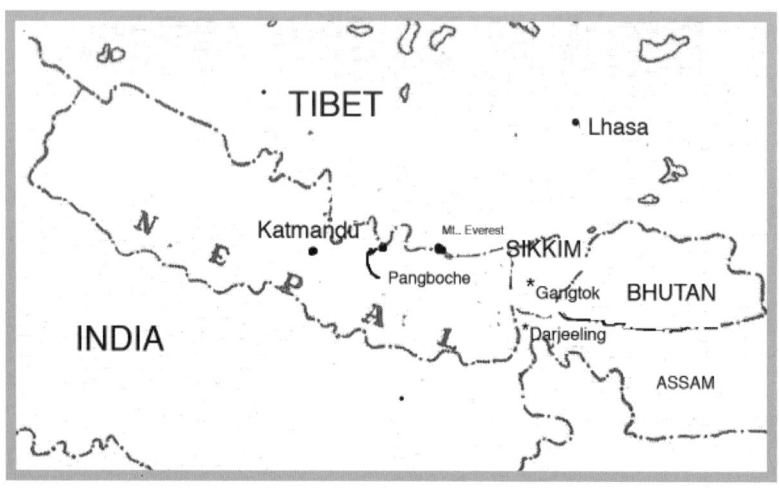

Nepal and surrounding countries.

Reports trickled in during the rest of the nineteenth and the beginning of the twentieth century. In 1921, for example, C.K. Howard Bury, leader of the Everest Reconnaissance Expedition, described enormous footprints left in the snow in Lhokpa Pass, at 6600 meters (20,000 feet). In 1925, British photographer N.A. Tombazi, a gentleman of impeccable reputation and a member of the Royal Geographical Society, observed a yeti-like creature crossing a grove of dwarf rhododendrons in the area of the Zemu Glacier, at an altitude of 5000 meters (16,000 feet).

These two examples will suffice for now, and we return to Peter Byrne who, in 1947, found employment with a London-based firm managing tea plantations in northern Bengal. Wrote Byrne:

> In those days companies were liberal with the amount of vacation time they allowed their young gentlemen assistants and just a year later I was able to take a month's leave of absence.[4]

In 1948, Peter drove to Gangtok, the capital of Sikkim, in his rusty Austin Seven. Accompanied by two Sherpas, he walked to the area of Green Lake, and after 10 solid days of trekking reached the Zemu Glacier. There, in the hard snow, he found a single footprint. The Sherpas, hardy mountaineers, reacted nervously and were eager to leave the area.

Famous mountaineers soon joined the roster of discoverers: Eric Shipton photographed a lone print on the Menlung Glacier in 1951; he was soon followed by John Hunt and Edmund Hillary. Prince Peter of Greece, a resident of Kalimpong, India and a respected anthropologist who pondered for many years the mystery of the yeti, was of the opinion that there was some factual evidence behind all those observations. Peter Byrne was often his guest.

Peter returned to Sikkim in 1956.

> By this time I had resigned from my tea company and gone into the big game hunting business in India. The life of a tea planter proved less attractive after Indian independence, and the big-game hunting safaris in India were in their infancy and just beginning to attract international attention. In the spring of that year, with time to spare between safari bookings, I once again set out for Sikkim. This time I went up the Sikkim–Nepal border route, a high ridge route that runs south to north and that climbs all the way into the Zemu area. I spent a month searching the frozen dwarf scrub up to 15,000 feet and found nothing more than wolves, snow leopards and bears, the common wildlife of the middle Himalaya.[5]

On his way back, trekking through cold and fog, Peter and his two Sherpas caught a glimpse of a group of men walking about 650 meters (2000 ft) below them. They aimed to join them: "a month in the hills leaves one wanting company." Soon, Peter and his companions were welcomed around the campfire where they met with old friends, one of them being Tenzing Norgay, Edmund Hillary's climbing mate. The men were members of the new Indian Mountaineering School created in Darjeeling. The conversation quickly turned to the yeti. As Byrne wrote:

> Tenzing told me that his father had encountered one which had crawled on top of his yak herder's hut one night and stayed there until driven off by smoke from the old man's yak dung fire.[6]

Tenzing also mentioned that he had recently spoken with an American, Tom Slick, who was planning an expedition to search for the yeti. Peter wrote to Slick in San Antonio, Texas; six months later they met in Delhi. Slick had long been interested in the mystery of the yeti. The photo taken by Shipton in 1951 during the Everest Reconnaissance Expedition had left just as vivid an impression on him as it had on Byrne. There was also that series of remarkable footprints photographed on the Menlung Glacier; the close-up of a footprint showing Shipton's ice axe for scale was unforgettable. These tracks were most enticing because of their resemblance to human footprints. For example, the big toe was larger and more human-like in shape than that of the great apes. Those striking black and white images kindled a passionate desire to resolve their mystery.

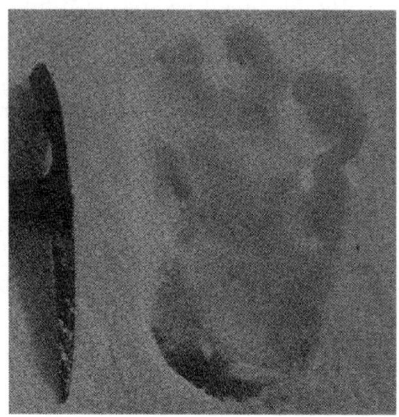

Few imagined that humans had made such tracks; even fewer supported the idea that they could have been made by a known great ape. Perhaps, as Odette Tchernine suggested, some unknown primitive hominid might have survived, living deep in the high-altitude forests or in unexplored ravines. Nothing could prevent it from venturing as high as 7200 meters (23,600 ft) where Shipton and his companions had discovered its tracks.

Shipton footprint, one of a series found by Eric Shipton and Michael Ward on the Menlung Glacier, 1951. PHOTO: International Society of Cryptozoology Newsletter, vol 5, No. 4.

In those days, people from all walks of life were taking an interest in the existence of the snowman. Mira Behn, for example, an English disciple of Mahatma Gandhi, was working for the Indian government, organizing cattle grazing at high altitudes in northern India. She listened carefully to reports by cowherds, who spoke of the *van manas* or *ban manas* (*van* or *ban*: forest; *manas*: man). She was concerned that should such an "unknown brother of the hills" be caught, care would be taken not to injure it.

One is reminded of the words of an old Indian quoted by Odette Tchernine:

> One day as I was walking on the mountainside, I saw at a distance what I thought to be a beast. As I came closer I saw it was a man. As I came closer still I found it was my brother.[7]

In some areas, such brothers are studiously avoided. In his *Himalayan Journals,* published in 1884 in two thick volumes, Dr. Joseph Hooker described (vol. 1, chapter 1) wild men living near the mountain passes of Sikkim, called the *harrum-mo* by the Lepcha. These wild men lived far from human habitation, spoke an unknown language, used bows and arrows, and ate snakes and other vermin. Lepcha guides did not dare approach them, let alone touch them.

In the middle of the nineteenth century, Bhutanese people believed in the presence of the snowman. They also thought that these creatures were becoming increasingly rare. According to them, there were three kinds of snowmen:

> Snowman Number One is large and fairly docile; the second type suggests a savage carnivore, about five feet tall, long-haired and apelike, and of muscular build. Thirdly, comes a "Little Man," shy and shaggy…According to the most recent statement I received from Bhutan, various types of Snowmen have common characteristics. The pungent smell and the high whistling, or mewing call, tally with some of the features attributed to other similar mystery creatures.[8]

Odette Tchernine points out that the snowman is Bhutan's national animal,to the point that in the 1960s it issued a series of postage stamps representing the various types of snowmen.

There was at that time general agreement between the various descriptions, profiling a creature closer to animal than to man, unclothed and without any weapons or tools, not having mastered fire, and incapable of articulate speech. Bhutanese touristic brochures featured the scenery, the flora and the fauna, including the snowman, which was treated with as much sobriety and interest as were more familiar instances of "reality."

Nearby, in China, there appeared in the English language *Peking Daily* of January 29, 1958, an eyewitness re-

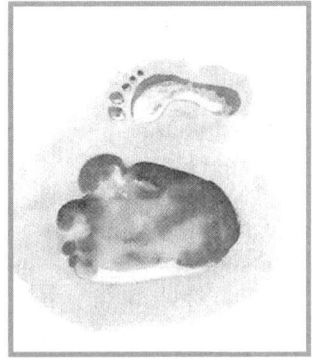

The lesser yeti and human prints compared. ILLUSTRATION: P. Coudray, *Guide des Animaux Cachés*, 2009.

port by Païn Hsin, a filmmaker with the People's Liberation Army. In 1954, returning from a shoot, he and three coworkers saw two "men," short and bent over, climbing, single file, a hillside in the Pamir Mountains. Païn Hsin and a cameraman followed tracks that closely resembled human footprints, except for their giant size. As night was falling, they had to turn back after about a mile (1.5 km).

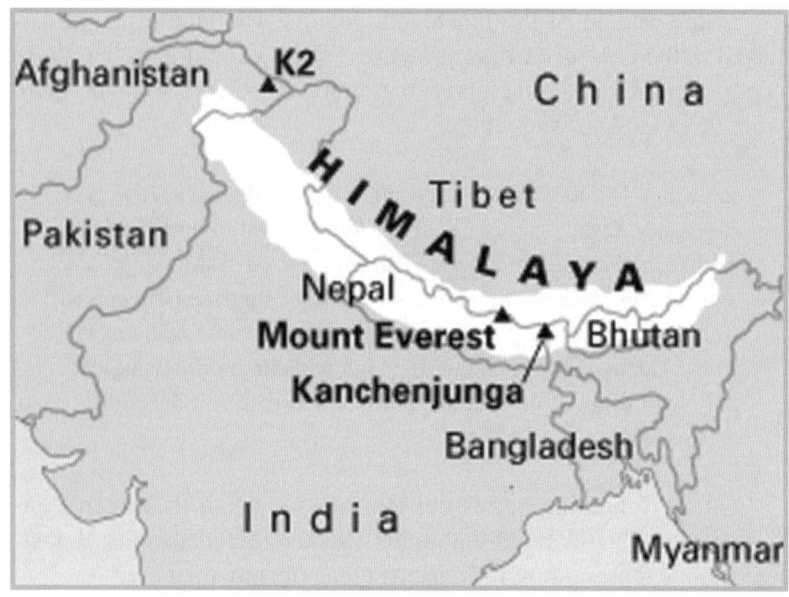

The Himalayan realm. PHOTO: Author's file

Païn Hsin had also lived in Sinkiang, a Chinese province just north of Tibet. He had stayed in the posts of border guards at the Russian frontier, in the Pamir Mountains. Once, the guards had taken away the carcass of a diseased (and deceased) cow some 40 meters away from their post. The meat must have been spoiled. At dawn, the guards saw a "wild man" sneaking away with pieces of meat. Whatever that creature might have been, Païn Hsin was sure of its existence in the Pamir wilderness.

On the Russian side, in the 1950s and 60s, researchers were looking for a snowman they called alma (or almasti) and were trying to map its geographical distribution, which turned out to resemble closely that of the snow leopard. It was suggested by some that it was the leopard's prints that were mistaken for those of the snowman. Others claimed that the reason that the snow leopard never attacks humans is that it lives in peace with the human-like yeti.

The proximity of snowman and leopard would of course suggest an answer to the recurring question: Why are the bones of snowmen never found? Simply because there are no leopard bones either. Predators totally clean every carcass. What's left is then swept away by mountain streams at snowmelt.

Thus, already in the 1950s there were serious research efforts focusing on the search for the snowman. Russian and Chinese scientists collaborated to create a map of its habitat in Russia and Asia. Witnesses came from all ranks of society: Sherpas, farmers, shepherds. Some had diplomas, like John Dalton Hooker, a physician; or Paï Hsin, a film producer working with soldiers familiar with the mountains.

The weight of credible reports, the plethora of tales, mythological accounts, scientific articles, and eyewitness descriptions by indigenous people all added up to a solid case for launching an expedition to prove the biological reality of the snowman.

---

1. Sherpas: A Nepalese ethnic group reputed for its skilled mountaineers. Lepchas: a Mongoloid people thought to be the original inhabitants of the mountains of Sikkim, where Tibetans and Nepalese also live.
2. Among the names describing "Surviving Hominids," Odette Tchernine finds the term *rakshasa*: mystical Sanskrit word for demon, reminiscent of *rakshi-bompo*: a hybrid Tibetan/Sanskrit word meaning powerful demon (cf Odette Tchernine, *In Pursuit of the Abominable Snowman*, p. 175).
3. Peter Byrne, *The Search for Bigfoot*, p. 96.
4. Peter Byrne, op. cit. p. 97.
5. Peter Byrne, op. cit. p. 97.
6. Peter Byrne, op. cit. p. 98.
7. Odette Tchernine, *In Pursuit of the Abominable Snowman*, p. 12.
8. Odette Tchernine, op. cit. p. 64.

2. Nepalese Expeditions

Confident in the authenticity of the information he had gathered about the snowman, Peter Byrne wrote to Texas cattle and oil entrepreneur Tom Slick, in San Antonio.[1] Slick flew to New Delhi to meet him. They traveled together to Katmandu to prepare for a trek on foot up the valley of the Arun into the mountains of northwestern Nepal.

At the end of three months Byrne found, at an altitude of 3300 meters (10,827 feet), some 25-centimeter-long (10-inch) five-toed tracks; Slick with a separate team also found a set of tracks, 32.5 centimeters (13 inches) long. After such encouraging beginnings, Slick decided to launch a more intensive search and spent the next six months gathering supplies. Another wealthy sponsor, Kirk Johnson, joined the team, which by now included five Europeans, 10 Sherpa guides and 65 porters. Well known naturalist Gerald Russell, who lived in France and had already traveled in the Himalayas with a *Daily Mail* expedition in 1954, led the expedition. The team also included photographer George Holton and German filmmaker Norman Dyhrenfurth. Peter Byrne and his brother Bryan were in charge of logistics. Pushkar Shumsher Jung Bahadur Rana, a retired Nepalese army captain, took care of liaison with local authorities. A carefully planned and well-outfitted enterprise!

The weekly magazine *Paris Match* followed the progress of the expedition. In those days only the cover page, some ads and a few photos were in color. Most of the pictures were in black and white, but their sharp contrast and the large format of the magazine left a lasting impact on readers' memories. On May 17, 1958, the front cover featured popular TV hostess Jacqueline Huet and a catchy hook: "Exclusive! The Hunt for the Abominable Snowman. On the tracks of the Yeti!"

Bryan Byrne demonstrating the use of the crossbow and anesthetic arrows.
PHOTO: *Paris-Match*

One of the full-page photos showed Bryan Byrne, surrounded by a group of villagers in local dress, handling a compressed air crossbow. It was hoped that the vial of curare in the arrowhead would put the "monster" to sleep. This brings to mind a conversation I had with Peter Byrne in Oregon in 1995 while he was showing me a stun pistol. Peter admitted that no one knew the appropriate dose for putting a bigfoot to sleep; the same is of course true for the yeti or any other unfamiliar creature. Too much can be lethal. How can one guess *a priori* how much curare it would take to cause only a temporary paralysis?

After leaving Katmandu, the explorers set up a base camp in the northeast valley of the Arun River. They visited Sherpa villages, some rather primitive:

> —it was a collection of a dozen stick huts hidden in the Himalayan forest. We approached, unarmed; the village was deserted. It was

only after about an hour that the inhabitants—about thirty altogether—gradually reappeared, one at a time. They had never seen white men…There couldn't be that much difference between their shacks and the Snowman's den. Their tools were fire-hardened sticks; their only weapon a long wooden spear and their bedding a pile of grass. However, the Sherpas understood their language. It was in that most simple of villages that we found ourselves again on the track of the Yeti.

As soon as we arrived, Gerald Russell, our lead hunter, interviewed the natives, with one of the porters acting as an interpreter. Each time the Sherpa pronounced the words *Metah-Kangmi*—abominable snowman—panic flared in the tribesmen's eyes. For untold generations the apemen were the terror of the valley. After an hour's discussion, an old man told us that a Metah-Kangmi had been seen three moons ago near a very small hamlet—two isolated hovels—down the stream."[2]

Gerald Russell, Peter Byrne, Norman Dyrhenfurth and, standing behind them, Bryan Byrne. PHOTO: Paris-Match

The nature of the terrain made getting around most difficult. The images gathered by filmmaker N. Dyhrenfurth give an idea of the nature of the ground. They also show the three mastiffs, Lou, Mary, and Blue, brought in from Arizona and trained for yeti hunting. The leaders of the expedition were all experienced mountaineers, familiar with Himalayan forests. Gerald Russell relied on solid principles:

- it was essential to remain friendly with the natives, otherwise, there was no hope of ever finding any kind of specimen;
- one had to be ready to get one's hands dirty and share in the work;
- one had to remain modest in behavior and appearance.

Many had trouble with one or the other of these conditions!

Camped a two-hour walk below the summit of a pass, in the thin mountain air, the men sought some protection from the wind, but the erection of a snow wall required great effort. An observer was always on guard, night and day. The explorers kept their guns handy: "our last defense against the yeti." Peter Byrne was sure that the snowman had been watching them for the last four nights. Should he be afraid? Not if he were to recall the adventure of Capt. d'Auvergne, a British soldier of French ancestry who, in 1938 in the Himalayas, was literally blinded by a violent blizzard. Lost and wandering, he faced death by hypothermia. A yeti carried him to his den and fed him. The captain recovered his composure and his sight, but the gray-furred creature had disappeared...

One night, Peter, Bryan and the photographer were awakened by the sound of footsteps. Struggling with the frozen tent zippers they rushed outside with their flashlights. Nothing. The creature returned the following night. "We were sleeping in our clothes, with the tent doors open. In thirty seconds we were outside. Not fast enough. We heard the snow squeaking uphill and a gigantic sigh. Our lights showed nothing but the darkness." (Peter Byrne)

In the daylight the explorers discovered 30-centimeter-long (12-inch) footprints; the creature had approached within two meters (6.5 feet) of Gerald Russell's tent. The big toe left a clear imprint. The tracks stretched over 200 meters (650 feet), as far as a rocky outcrop. Peter dispatched two Sherpas with the news and readied enough supplies for a sweep of the area over the next two weeks. Most of the materials were left in a cave used as observation point.

After four months the expedition members left Nepal, called elsewhere by various commitments. Peter and Bryan Byrne continued their search alone for five more months.

Finally, when the snow became too deep, the brothers walked back to Katmandu. They consumed the last of their supplies and found that much of their equipment needed replacing. They spent six restful weeks in the modest Royal Hotel before a message from Tom Slick requested them to return to the freezing summits to photograph a yeti. A week before Christmas the two young men left behind the comfort and amenities of the city. Peter noted that they were so taken by what Slick called "the ultimate quest" that during the first year they worked for nothing. In the second year, they accepted wages of $100 per month each.

The two men traveled light, without tents or food supplies. They decided to live like mountain shepherds, sleeping in wooden shelters or in caves when they climbed above the tree line. They spent another nine months in northeast Nepal.

Did they perhaps find time to read in the papers the report of a famous Russian geographer, Prof. S.V. Obruchev? According to him, the yeti was as tall as a man, covered with brown or gray hair or fur. Its feet were 25 to 35 centimeters (11–16 inches) long and wider than a human foot; the big toe, as well as the next toe, was well separated from the other toes. The prints were reminiscent of both those of a man and of a great ape, similar to those of a Neanderthal but more primitive. The snowman fed on plants, herbs, berries, roots, insects and small mammals.

During her research on the northern slopes of the Caucasus, Dr. Marie-Jeanne Koffmann and her team had discovered two dens of almas:

> They found a heaped-up larder consisting of two pumpkins, eight potatoes, a half-chewed corn-cob, two-thirds of a sunflower centre, blackberries, and the remains of three apples. Mixed up in this hoard were four round pellets of horse dung. It seems the Almas are very fond of this substance because of its salt content.[3]

Peter Byrne reports that an unknown primate, probably a yeti,

Bernard Heuvelmans and Edmund Hillary examining a yeti scalp. PHOTO: Author's file

living in a canyon of the Arun River also had frogs on his menu; at least that's what the natives said, since Peter and Bryan, in spite of the hundreds of kilometers of travel in the high mountains, never met the snowman.

Nevertheless, some findings suggested that the brothers had found solid evidence: first, the discovery of the scalp of a yeti in a monastery in Pangboche; later (1960), Sir Edmund Hillary, the conqueror of Everest, brought back an identical imitation of the Pangboche scalp from an expedition in search of the snowman. Peter and Bryan also found in Pangboche what they took as the mummified hand of a yeti. James Stewart, the actor, and his wife agreed to smuggle it to London in their baggage.

For three years in the Himalayas the Byrne brothers enjoyed a life fully as adventurous, and sometimes as dangerous, as young men can possibly imagine.

The Pangboche hand was perhaps an essential element of a sacred ritual. In his quest for knowledge, Peter Byrne interfered with this relic, a supporting artifact of a holy prayer to the emanation (from the Latin *manus* = hand) or even the manifestation of the Supreme Being. As a result, the affair of the Pangboche hand has taken on a divergent mythical status. This ritual Tibetan object has now entered the domain of western science, passing from one mythical realm to another. In London, W.C. Osman Hill conducted a physical examination of the pieces (thumb and a finger bone) that Byrne supplied. His first findings were that it was hominid, and later in 1960 he decided that the Pangboche fragments were a closer match to a Neanderthal. However, more recent DNA testing has shown the bones to be of human origin.[4]

The yeti scalps preserved as sacred relics in Tibetan monasteries had strictly religious purposes. The specimen brought back by Sir Edmund Hillary and examined by Bernard Heuvelmans turned out to be an artifact constructed from the skull, skin and hair of the serow goat (*Capricornis thar*).[5] The lamas wore that cap during ceremonies in order to represent the snowman.

According to Bernard Heuvelmans, Tom Slick's most important contribution was to distinguish between two kinds of yeti:

- one was covered with black hair and stood up to 2.40 meters (8 ft) tall;
- the other was smaller and with a reddish pelt.

The former lived in Tibet, northern Sikkim and Nepal, at altitudes above 4000 meters (13,000 feet).

Bernard Heuvelmans named it *Dinanthropoides nivalis*, the "abominable snow anthropoïd." Should it be discovered some day that its teeth are similar to those of a *Gigantopithecus*, it may become necessary to rename it *Gigantopithecus nivalis*, to distinguish it from the giant primate of the late Pleistocene (500,000 years ago). As *Gigantopithecus* is a member of the Pongid family—to which also belongs the orangutan—so is *Dinanthropoides* a member of a parallel branch of the same family.

The yeti is thought to be omnivorous, feeding on roots, bamboo, fruits, insects, lizards, birds, small rodents and even larger prey such as yaks.

In 1957, a Nepalese paper described the terrifying massacre in 1922 of 31 soldiers in an isolated village 80 kilometers (50 miles) northeast of Katmandu. Of the 32-member unit marching at night toward the Tibetan border, a single man had survived. The survivor returned with a 10-man-strong squad. The monster was found asleep after having devoured many of its victims. It took at least two volleys to kill it. The squad leader kept the head. However, a quarter century later no one could find it.

The smaller yeti, stocky and hairy and no taller than a 14-year-old child, has a pointed head with thick hair and moves quickly. He eats mostly picas and other small rodents. He is said to be more timid than aggressive and appears quite intelligent.

There are many reports about these two creatures. One of the most credible comes from Commander E.B. Beauman of the RAF, who found footprints on a glacier at 4200 meters (13,600 feet) during a mountaineering expedition. However, those expeditions specifically devoted to finding the yeti never managed to catch a glimpse from near or far—only prints, hair and droppings. From an examination of the yeti's feces, in which he found hair, bones, the whiskers of three mice, a feather probably from a baby grouse, as well as grass, thorns and the claw of a large insect, Gerald Russell concluded that it must be omnivorous.

A large group on an organized expedition rarely manages to observe a shy animal. In 1950, in the hills, 30 porters were needed to support three explorers. The *Daily Mail* expedition of 1954 counted 300 men. Journalist Ralph Izzard commented that such an assembly stands out like a line of cockroaches on a tablecloth! It might be more effective to split into small groups hidden for longer periods in strategically selected places. Bernard Heuvelmans also did not think much of quick expeditions: animals do not have time to become used to the presence of the intruders and the explorers are there too briefly to merge into the background, which of course takes some time.

---

1 "Tom Slick, the king of Texas oil, who reads only Kipling and whose only hobby is adventure." Peter Byrne in *Paris Match,* no. 475, 17 May 1958.
2 Peter Byrne, *Paris Match,* loc. cit. p.17.
3 O. Tchernine, op.cit. p.21.
4 The hand and the scalp were stolen from the Pangboche monastery in the 1990s. Recently (spring 2011), New Zealander Mike Allsop had replicas made, which he plans to return to the monastery "to help them have an income

again," attracting tourists to view the relics. (www.bigfootencounters.com/articles/yeti-hand.htm) The hand bones (a thumb and a finger) were recently rediscovered at the British Royal College of Surgeons; Dr. Rob Ogden of the Royal Zoological Society of Scotland performed DNA analysis, and the results were revealed on the BBC on December 27, 2011. (www.bbc.co.uk/news/uk-scotland-edinburgh-east-fife-16316397).

5   Anthropologist Myra Shackley disagreed with this conclusion on the grounds that the hairs from the scalp looked distinctly monkeylike and contained parasitic mites of a species different from that recovered from the serow.

3. The Book of Small People

"The Book of Small People" is the literal rendition of the Chinese ideogram for a "comic book." One of the most famous authors in this genre was Hergé, the Belgian creator of Tintin, the young reporter, and his dog Snowy. *Tintin in Tibet* is a classic graphic novel adventure. Bernard Heuvelmans was a friend of Hergé and his technical advisor on the yeti. In the story, Tintin's young Chinese friend Chang was flying to Katmandu when his plane, caught in a storm, crashes in the mountains. Convinced that his friend has survived, Tintin goes looking for him and eventually finds him in a cave, where he has been sheltered and fed herbs, roots and small animals by a greater yeti.

Is the yeti, as drawn by Hergé, a real plantigrade biped? Is it an anthropoid great ape? Hergé must surely have pondered such questions after reading Heuvelmans' description of the footprints, with the big toe close to the other toes, as it is in humans. Besides, Heuvelmans added, anthropoids are usually quadripedal; even when it stands up, the orangutan uses its arms as crutches. Further still according to Heuvelmans, the great apes are not built to walk on their hind legs. The position of their skull and the shape of their spine make them lean forward: the gibbon, for example, stands up by using his long arms for balance.

Nevertheless, great apes have been seen walking on their hind legs, if only for a short time. Dr. Sydney Britton, a professor of anthropology at the University of Virginia, observed the behavior of a chimpanzee recently arrived from Africa. It had just snowed and the animal stood up on its hind legs. Dr. Britton speculated that snow had played a crucial role in human adaptation to the bipedal stance. Similarly, in the Himalayas, a great ape could have learned to walk on its hind legs to minimize the area of skin in contact with the frigid snow.

Humans have kept the plantigrade feet of primitive mammals. Human feet are not derived from the prehensile feet of tree-climbing monkeys. Evolution is to be thought of as having progressed in the opposite direction: the feet of the great apes evolved from those of primitive humanoid ancestors. The great ape's foot is "well developed," with a strong heel, long and curved toes and, of course, a big toe that moved apart and became opposable. Such a big toe turned out to be useful to climb trees. However, a lineage of creatures that evolved to a larger size would stop climbing trees. This reasoning has led some experts to speculate that the evolution of the gorilla has by now come to an end.

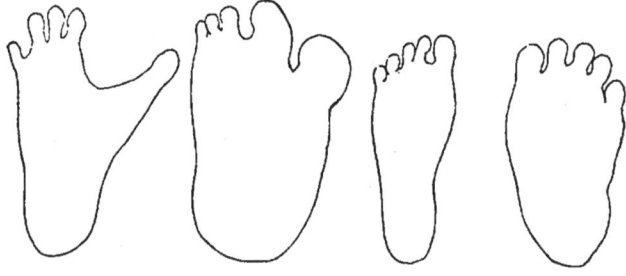

Left to right: feet of gorilla, snowman, man, bear. ILLUSTRATION: Author's file.

A giant great ape would be more comfortable in a relatively open area: mountains would become an ideal habitat and the snow cover would lead it towards bipedia. This is why Heuvelmans speculates that a race of giant apes, sporting primitive plantigrade feet and bipedal posture, could have developed in a mountainous domain.[1] As Heuvelmans remarks, there already exists in the mountains a race of giant carnivores possessing these very characteristics: the bears!

In June 1942, Slavomir Rawicz and three other exhausted wanderers were rescued in Sikkim by Indian Army soldiers. They had walked across Mongolia, Sinkiang and Tibet, a total of 3000 kilometers (1900 miles). There were originally seven of them who had escaped from a Soviet labor camp in Siberia. Three of them died on the way. During their journey (May 1942) the four survivors said they had met a couple of yetis in a Tien Shan pass between Tibet and Nepal; they watched them for two hours[2]:

> They stood upright, sticking out their mighty chest, with their hands, at the end of excessively long arms, hanging at the level of their knees. Their ears were flat against the skull and, seen from behind, the side of their head showed a straight line from the top of the head to the shoulders. One was reminded of a "Prussian nape." I couldn't fit this animal in any category; I finally opted for a cross between a bear and an orangutan.[3]

Bernard Heuvelmans readily dismissed the idea of a bear and thought that many details strongly suggested an anthropoid ape. He also remarked on the military stiff-necked appearance, sometimes found in older gorillas: "The striking rigidity of the yeti's neck is likely to be linked to a long-established bipedal stance."[4]

Although Heuvelmans' hypothesis appears somewhat at odds with more established ideas, it is instructive to follow his reasoning, and to see the elements brought in by a (crypto)zoologist to understand an unfamiliar animal. Hergé, the cartoonist, made good use of the features suggested by his friend Heuvelmans: general appearance, footprints, pelt, sagittal crest, long arms, small prey judging from the size of the bones in the den. The snowman's hair also agrees with Heuvelmans' description.

Sheltered in a monastery, Tintin and his friends are guided by a lama's vision towards the area where the snowman picked up young Chang. The monks know about the *mi-gou,* as Tibetans call the animal named yeti in Nepal. However, in spite of the fears expressed by the monks, the mi-gou acted in a most friendly fashion towards Chang. Freed unharmed by his friends, Chang remarks:

> it behaved towards me in such a way that I often asked myself if it was not a human being…

In striking contrast, Odette Tchernine speaks of the terrifying story told to her by a journalist friend about a little girl kidnapped by a yeti. It was before the Second World War; the child was never seen again. Does *Dinopithecus nivalis,* the "terrible snow ape" merit such a reputation? We shall soon return to this question.

Odette Tchernine emphasized the place given to wild men in Tibetan monasteries, near which footprints are often found:

> The lamas in these *gompas,* as those monasteries are called, beat gongs to scare them away. An account describes how some yeti were given drink and then murdered as they lay drunk within the gompas' enclosure.[5]

Coincidentally, according to Tintin's Sherpa cook, the yeti loves beer to the point of falling asleep drunk. On that day, he said, the villagers had tied him up, but when he woke up he broke his bonds and fled. Hergé drew freely from the stories told by Himalayan villagers.

Many authors and travelers speak of the relations between the yeti and people. American novelist and non-fiction writer Peter Matthiessen was on an expedition to study the snow leopard, sitting around the campfire with his Sherpa companions, when he alluded to the yeti. Their view was that it was more a man-creature than an ape-creature. It was not dangerous, but to meet it brought bad luck. Their grandfather said that the yeti stole from the crops and had in the past been killed with poisoned barley. The Sherpas also thought that the yeti was a Buddhist. Tibetans claim to be the descendants of a monkey-god, and a Sherpa legend says that a monkey that became a Buddhist hermit lived in the mountains and married a she-demon; their children were *mi-the,* or yetis.

Matthiessen alluded to the animist gods of pre-Buddhist religions whose focal point was the mystery of the *sangbai-dagpo,* or hidden lords. That religion, the predecessor of lama Buddhism, was obsessed by the transmigration of the human soul into the bodies of lower anthropoids. Members of that sect venerate the "abominable snowmen," and the head, feet and hands of dead specimens play a role in their rituals. One should not underestimate the influence of that animist doctrine on Tibetan Buddhism; it also leads local people to protect yetis from the Europeans' probing investigations.

Dr. Charles Stonor, an anthropologist who was the associate director of the London zoo, had traveled in Tibet in 1953 before joining the *Daily Mail* expedition in 1954. It was he who quoted the words of a lama of the Pangboche monastery, according to whom the remains of the snowman were the object of a special cult in many monasteries. The monk added that he had actually seen two snowmen in Tibet.

Within the Pangboche Gompa, Stonor discovered a relic which soon became famous: the scalp of a yeti, three-and-a-half centuries old. In Heuvelmans' view, the shape of the head suggests a personality quite different from that of human being's: "Only the examination of an actual specimen of the Snow Man will allow verification of my conclusions and the completion of a preliminary description of Dinopithecus."

These words written in 1955 are still true today. The analysis of the "relic" revealed it to be a fabrication without scientific value (it was made from the skin of the serow, a type of goat). However it was one of three such scalps, discovered at that time or later. The others have not been analyzed. The biological nature of snowman remains as elusive as ever.

From a symbolic perspective, however, the scalp or the skull suggests the *seat of consciousness*, the hand is the *emanation* of the spirit, and the footprint, or rather a series of ephemeral tracks in the snow, speaks of the *path* followed by these beings, the way forwards. In a mystical context, the trilogy footprint/hand/scalp takes a familiar meaning for Tibetan Buddhists: it is a call to go further, higher, towards the awakening, the Enlightenment, the ultimate stage of human evolution.

Meditation leads to enlightening intuition, towards the insights that break through the veil of illusion. The inner vision is the way that leads from the world of appearances to that of deep truth beyond the world of mere phenomena.

Until recently, sometime before the definitive annexation of Tibet by

Portrait of the yeti.
ILLUSTRATION: *Science et Avenir*, 1958

China in 1965, generations of hermits followed each other high on the slopes of the Himalayas, at 5000 meters (16,000 feet), where Milarepa sat and meditated, and where the yeti spends its life. Mila of the cotton robe, his only garment, was an ascetic and a poet:

> Within the ocean of transmigration between the three worlds, the mystical body is the great fisherman. As long as one is concerned with food and clothing, there can be no giving up the material world.[6]

Those journalists, explorers and scientists searching for the yeti in the 1950s were probably unaware even of the rudiments of Tibetan mysticism, which must be approached with patience and humility. They might have avoided some disappointments if they had adopted the explanation once offered by a Korean wise man to globetrotter Alexandra David-Néel: We create the world by our thought, but the world also creates us, including our thoughts. There is interdependence. The world (*samsãra*) does not exist without us; we do not exist without samsãra.

Two Sherpas.
PHOTO: *Paris-Match*

1. Bernard Heuvelmans, *On the Track of Unknown Animals*, p 168–171.
2. There are claims that Rawicz did not make the journey as stated. It is said that he related the experiences of another man, Witold Glinski, and fabricated the yeti sighting. This, however, has not been substantiated other than by testimony.
3. Statement by Slavomir Rawicz, as quoted by Bernard Heuvelmans in *On the Track of Unknown Animals*, p. 183 ff. See also Slawomir Rawicz (with Ronald Downing), The Long Walk (1955).
4. B. Heuvelmans, op. cit. p. 185.
5. Odette Tchernine, op. cit. p. 68.
6. Encyclopédie des Mystiques, tome 4, p. 28.

4. The Yeti and Ethnomedicine

The following sections are an attempt to place the presence of the yeti within a broader framework. A deeper perspective will help demonstrate that the yeti, and more generally the wild man, as well as the rituals that surround them, are not specific to the Himalayas but find their counterparts in remote areas and ancient times.

A sidebar in a French magazine recently caught my eye; it described how some chimpanzees took care of their health. While in Uganda, Sabrina Krief, a primatologist with the CNRS, noted that a female chimpanzee suffered from digestive problems. She saw her draw apart from the family group and gather the bark of albizia, a shrub usually of no particular significance to the apes. Two days later, the female chimp was better. An analysis of her feces showed her to be free of the microorganisms responsible for the condition. Dr. Krief remarked:

> The observation of animals is perhaps a new approach to discover tomorrow's medicinal plants.[1]

According to Haida sculptor Ralph Bennett, the first shaman, guided by his intuition, followed a sick bear into the forest. He saw it grab some pieces of bark and chew them. The shaman filled his medicine bag with this bark and experimented with it on himself.

Thus, was empirical medicine born. It has flourished through the centuries. Roman civilization, ahead of its time in many areas, kept intact ancient traditions, as attested by numerous authors. Seneca, writing in the first century after Christ, thought, "the medicine of yore was the science of a few herbs used to stanch blood and help the healing of wounds." Seneca was of course speaking of ancient times, back at the beginnings of Rome (seventh century BC). Perhaps in those days physical labor and simple food were enough to guarantee good health, without much need for physicians. The poet Lucretius believed that the first men were not subject to diseases. However, the absence of physicians does not imply absence of a medicine based on simple remedies. Cato the Elder held cabbage as the most important and healed himself with a regimen of wine and cabbage; he also relied on incantations to cure dislocations!

There is no doubt that herbs played an important role in curing the ills of both men and cattle. As cities expanded (Rome had a million inhabitants by the first century AD) so did the scope of medicine; Seneca commented: "as the number of dishes increase, so do the maladies."

Pliny the Elder, (first century AD) deplored, as did many others, the fact that his contemporaries had abandoned the simple rustic life of their ancestors. His reference to "The innocent wisdom of early days"[2] is a reminder of the old myth of the Golden Age, before man was corrupted by civilization.

Today's renewed interest in ethnomedicine and ethnopharmacy reaches back to the most ancient practices, modified and skewed by superstitions well before they were replaced by modern science. Science in turn, victim of its own rigor and specialization, no longer fully satisfies a large segment of the public. Ancient healing practices are being viewed from a new perspective. It has long been the custom for medical practitioners to heal the spirit as well as the body of the patient. Today, there is a rediscovery of traditional therapeutic methods that treat the sickness by invoking an invisible world accessible to the initiated practitioner, be he shaman, healer, medicine man or medium. Such healers use man-made tools: masks, drums, rattles or costumes.

One cannot overemphasize the importance of ritual in healing practices. In his book, *The Hero with a Thousand Faces*, mythologist Joseph Campbell provides numerous examples. For instance, the !Kung Bushmen of South Africa, whose main musical instrument is their own body, move their feet and their hands with great virtuosity to create extremely complicated rhythms. Behind the drums rise the many voices of healing chants. During these healing dances, the Bushmen commune with each other more deeply than ever, to such a degree that they become a single body. Within this close-knit configuration, they come face-to-face with the gods.

Such practices take us back to earlier eras, such as the Stone Age, which stretches over most of the Quarternary and consists of three periods:

- Paleolithic: from the Greek *palaios* = ancient, and *lithos* = stone
- Mesolithic: from *mesos* = middle
- Neolithic: from *neos* = new

Joseph Campbell drew attention to the famous wall painting in the Grotte des Trois Frères in southwestern France, dating from about 15,000 BC. It features the "dancing sorcerer," a shaman dressed in animal skins, playing some kind of musical instrument, probably a musical bow.

The dancing sorcerer, Grotte des Trois Frères. PHOTO: Author's file

In those days, the European continent looked like today's Siberia. It was an immense frozen tundra through which wandered large herds of animals—bison, rhinoceroses, mammoths—as well as our human ancestors. According to the experts, they lived in small bands of hunters, spending summers in outdoor camps, winters in caves, where they left mysterious documents in the form of cave paintings. Campbell thought that the caves were used as percussion instruments. Sound produced by striking stalactites was amplified by the acoustic properties of large caves.

Soon after the magnificent dancing sorcerer was painted, a significant change in climate took place: the end of the ice age. The glaciers melted; some animals, like the mammoth and the woolly rhino died off; others migrated eastwards, some crossing over to Beringia before the flooding of Bering Strait.

The Mesolithic is characterized by the appearance of agriculture and the husbandry of goats and cattle (10,000–2,700 BC). It is in the Neolithic that skin drums became common.

American musician Mickey Hart, who wrote a study on drumming through the ages, considers that the individual who discovered the percussive power of stretched skin was a genius.

The drum plays a very important role in the work of the shaman. Its rhythms carry him, as if on wings, to the higher dimensions and, during the trance, allow him to work miraculous deeds. He flies like a bird in the world above, or he descends into the nether world as a reindeer, a bull or a bear.

> Among the Buriats, the animal or bird that protects the shaman is called *khubilgan,* which means metamorphosis, from the verb *khubilku,* to change one self, take another form.[3]

However, the flight of the shaman is strewn with obstacles. The Buriats tell how their first shaman was able to bring the souls back from the kingdom of death. Following a complaint from the Lord of Death, the Great God in Heaven decided to test the shaman. He caught the soul of a human and shut it into a bottle, stopping the spout with his thumb.

> The shaman searched through the forests, the rivers, the mountains, even the land of the dead. Finally, astride his drum, he climbed all the way up to the upper world…[4]

Having watched the Great God in Heaven, he shape shifted into a wasp, stinging him in the forehead. Surprised, the Great God removed his thumb and the prisoner fled from the bottle. In revenge, the Great God broke the shaman's drum, thus diminishing his powers. The punishment could have been worse!

Incidentally, during the fourteenth century, there used to exist in the area of Toulouse, in France, a middleman, the *armier,* who transmitted messages from the dead. Of course, the bishops frowned on this parallel clergy. In the mountain villages of the Causse district, such armiers, or *âmiers* (from *âme* = soul), continued to play their role of intermediaries between the living and their deceased parents until around 1900.

For Campbell, there is no shaman without a drum. The object is endowed with specific properties. As with any drum, its tone depends on who was its past master. The instrument certainly often has a strange influence. Mickey Hart relates how a skull drum—*thod raga* in Tibetan—came into his hands. The gift from a friend, the antique double drum was constructed from the skulls of two sisters who died in an epidemic. In India, such an instrument is called a *damaru*.

Monk holding a damaru and a bone flute. PHOTO: Author's file

Although not very large—a skull fits in the palm of the hand—the drum had a powerful sound. Soon, Hart suffered from acute nausea. Further mishaps followed: he dropped things, fell, hurt himself; he narrowly escaped death when his car rolled into a ravine. Hart consulted a Buddhist monk, Tarthang Tulku,[5] who told him that by playing a damaru he had accidentally brought something back to life. Fortunately for him, there then followed a luckier period, characterized by a revival of his own creative forces, long deadened, which led to a series of new musical compositions.[6]

Hart understood that the powerful drum was more than just a simple rhythm machine. Playing it had opened a door and had, at the beginning, freed some dangerous spirits. A transformation then took place through an intimate and fruitful chemistry. It seemed that Hart had been transported by ascending forces after having stagnated and nearly floundered into a potentially fatal abyss. The new path that opened to him was a call for artistic exploration of the magic of rhythm, he being first and mainly a musician.

Again, the shaman is first and foremost a healer. He sometimes relies on hallucinogens: in America, mainly tobacco or mushrooms; in Tibet, powerful incense fumigations. In appropriate doses, well known to healers, some plants, such as the iboga of Gabon, have highly beneficial effects. Its users call iboga the herb of wakefulness. For the non-initiated, the trip may, however, be short and painful. The presence of a master is absolutely necessary. The shaman himself may even get lost in the trance. He lets his soul leave his body to travel beyond. But he is supposed to control that excursion in the other, higher world, a trip destined to heal the patient, who also experiences the trance while remaining passive. In contrast, the shaman's trance is active. An author wrote, "the shaman is a visionary and adventurer of the great beyond," concluding, "clinical descriptions can't account for the extraordinary wealth of the shamanic imagination, perhaps the universal embryonic form of artistic creation."[7]

North American witch doctor, 1590.
ILLUSTRATION: public domain

However, the dangers are many and go far beyond personal risk, as illustrated by the story of the tree with two branches, one bearing delicious fruits, the other poisonous ones. During a famine, a villager climbs the tree and absentmindedly picks a fruit. Luckily, it is an edible one. The villager is complimented and it is resolved to cut the other branch. Alas, the tree soon dies.

Might this perhaps suggest that western science, just like the tree with only one branch, lags behind the shamanistic pharmacopeia?

Let's now draw nearer to those regions of current interest. Take for example the Magars, an important Nepalese ethnic group: 1,623,000 people, i.e., 7.1 percent of the total population according to the 2001 census. To this day, these hill people from northern Nepal still organize their religious life around a community of shamans. According to Anne de Sales, when a shaman starts a song, he pays homage to his ancestors. It is as if he were saying, "I was born from your dances, I was born from your drumming."[8]

The Magars speak of a primordial Golden Age contrasting with the Blind Age, the dark period in which we now live. The eras that followed the Golden Age became increasingly more chaotic until the first shaman appeared and discovered that it was evil spirits and sorcery that were at the back of sickness and misery.

Today, each shamanic initiation and each healing process hark back to that primordial Golden Age. The blending of the sacred time and of the present, of myth and reality, keep this tradition alive.

---

1. "Profession: traqueur de molécules," article by Amélie Padioleau in *Notre Temps,* no. 420, décembre 2004. The CNRS is the Centre National pour la Recherche Scientifique, the principal French governmental research organization.
2. Jacques André, *Etre médecin à Rome*, p. 17.
3. Joseph Campbell, *The Flight of the White Gander*, p. 166.
4. Joseph Campbell, *The Hero with a Thousand Faces*, p. 199.
5. Tarthang Tulku, born in 1934, left Tibet when he was 25, following the occupation of his country by China. After teaching in India for 10 years, he settled in Berkeley, California, where he founded a number of associations (welcoming Tibetan refugees, preservation of Tibetan culture). He is the author of a dozen books on Buddhism.
6. Cf. the Grateful Dead album, *Blues for Allah,* 1975; also Mickey Hart, *Voyage dans la magie des rythmes* [Drumming at the Edge of Magic: A Journey into the Spirit of Percussion], 1998
7. Luc de Heusch, "Possédés somnambuliques, chamans et hallucinés" in *La Transe et l'Hypnose* (ouvrage collectif), p.42, Imago, Paris, 1995.
8. Cf. Anne de Sales, *Je suis né de vos jeux de tambours,* as well as Holger Kalweit, *Shamans, healers and medicine men.*

5. The In-between Times

Before continuing our exploration of the sacred time and of the present time, let's set a few reference points. Traveling deep into the Asiatic continent in search of the yeti, we encounter places, things and people with strange names, as well as unfamiliar rituals. Let's start with a few ideas linked to practices once current in the West. The vocabulary will now be as familiar as the cultural background with which we are familiar. However, whatever the scenery, the phenomena described are essentially the same as those less familiar traditional celebrations in the most remote areas (the Himalayas, the Caucasus, China…).

Let's start with a custom still very much alive in Europe: that of the Yule log, best understood through British traditions. The term Yule log comes from Scandinavia, where the winter solstice was celebrated on the feast of Yule. Yuletide, as a synonym of Christmas, was nearly abolished by the Puritans who associated it with the pagan Saturnalia, a feast the Romans celebrated with debauched revelry in December. Orgies were the most spectacular aspect of these holidays, during which everything seemed permitted: men and women went as far as exchanging clothes, a masquerade viewed as a regression to chaos. In these words:

> Disguise, I see, thou art a wickedness,
> Wherein the pregnant enemy does much.[1]

Shakespeare expresses the views of the seventeenth century Puritans, which prevailed until the celebrations, now firmly linked to the Nativity and Christmas, were again encouraged in the reign of Charles II during the English restoration. The Twelfth Night concludes the Christmas season with merrymaking, heralding the Epiphany, when the three Wise Men, the Kings of Orient, arrive in Bethlehem after a 12-day trip and offer incense, myrrh and gold to the child Jesus. Shakespeare's play by that name was written as an entertainment for the Christmas season.

The Christmas carol, *The Twelve Days of Christmas,* speaks of a different gift every day. The number 12 becomes particularly important when one recalls that most ancient people used a lunar calendar. A lunar month is on the average 29.53 days long, with 12 months adding up to 354 days. Eleven or 12 days have to be added to make up a solar year of 365 days and a quarter.

Additional days were inserted at different times depending on the particular calendar. The Egyptians, using 30-day months, only had to add five intercalary days to make up a year. The Gregorian calendar has retained one lunar month, February, which for a long time was the last month of the Roman year. With months of unequal length, the spiritual meaning of the inserted days has been lost; the additional days, spread from month to month, have become just ordinary days. It is difficult today to imagine the state of mind that accompanied those special days. The carnival period has preserved some of the flavor of these periods of feast and chaos, a time when the kings of clowns sit on the real king's throne.

Today, instead of the shamans and initiates of old, spirits and demons are represented by wearers of masks and costumes, often burlesque characters. *Gody*, the ancient Slavic and German New Year period, coincident with Christmas and the solstice, has left strong traces in today's customs. In Eastern Europe, such rituals have retained their vitality and embody a perspective older than Christianity. As darkness of winter overcomes the light, as the

Gody mask, Poland, 2004.
PHOTO: Author's file

wind roars and spreads its frozen wings, the souls of the departed approach the windows, begging for a new life. Villagers huddle by the fireplace, fearful of lurking underground spirits transformed into animals such as the wolf, or the bear... During Gody, spirits appear, attracted by the light and come to talk to the living, who in turn fear them, for spirits know the future and can influence man's fate.

The spirits are befriended through offerings of oats, straw, hay or bread. The communal banquet is followed by the distribution of the 12 gifts. Conjuring spells—against the wolf or the bear—are spoken in a language that has become incomprehensible. In the street, as during carnival, masked youth pursue the maidens, sometimes tying them up to carts or plows. Most often they are made to draw small carts, or are pulled around sitting in such carts. All this hazing aims at transforming the woman in a mare-goddess, the divine horse mother, supreme deity of yore. *Epona,* the Gaul goddess, for example, means Great Mare.

These rites bring us back to those "in-between times," the additional days. According to religious historian Mircea Eliade, such interruptions in the flow of time ensure the cyclic renewal of time. Each year, the world is created anew. As a new creation emerges, the living dead rise, as there is no longer any barrier between the dead and the living. Time is suspended and the deceased hope for a return to life.

The orgy signals a reversal of values; social norms are dissolved, the king is subjected to ritual humiliation by taking the place of the slave; there is regression. It is a kind of flood that precedes the regeneration, the second birth.

In some societies, rituals of extinction and rekindling of the fire dominate; in others, it is the expulsion of demons, through noise, dances and violent gestures; elsewhere, the scapegoat is expelled, either in human (as a wild man) or animal form.

What really matters is the absolute erasure of all sins and faults of the society as a whole, more than just purification. The past year is done away with, fully spent; all ashes are removed from the hearth. The time has come to restore harmony. A log that will burn brightly is carefully selected. The Yule log is a synonym of a new cycle. The creation of the world is re-enacted with the participation of the members of the community. During the "in-between time" they have lived in mythical time. Novelist and mythologist Jean-Charles Pichon speaks of the Saturnalia as "the Freedom of December."² Sadly, the citizens of the rich countries have forgotten the meaning of ancient feasts and this expression makes little sense for them.

---

1   William Shakespeare, *Twelfth Night* (II, 2).
2   That is the title of one of Pichon's first novels, *La Liberté de Décembre*, (éditions Ariane, 1947). The irreversible disappearance of ancient traditions leaves a vacuum, an expectation, most keenly felt by young people. Jean-Charles Pichon writes: "The street gang is an effort at returning to the Great Secret. We recognize in these spontaneous groupings many of the characteristics of vanished cultures: emblems, totems, initiation, clearly emerging as an attempt to create a society independent from the world of adults, irrational perhaps, but easier to live with." (*L'Homme et les Dieux*, p. 492).

6. Dasaï, Dasain or Dashain

Traditional holidays still perpetuate ancient significance in many countries. Such a celebration is the Nepalese Dashain held over a fortnight during September and October.

At the entrance of each village, a large swing is erected—a see-saw between the past and the present. Houses are cleaned through and through; banquets are prepared. There are dances, drink-ups, card games played for money. Kites, their strings lined with glue and broken glass, are launched in aerial combat: the goal is to cut the opponent's string. All this to celebrate the nine faces of the goddess Durga, whose role is to extinguish the forces of evil, symbolized by the demon Mahisasur, in the shape of a water buffalo.

Barley seeds are planted in pots of holy water and dung, called *kalash*. After a few days, barley sprouts appear: the sacred herb, called *jamara*, a sign of the bountiful influence of the goddess Durga. On the seventh day, the royal kalash is taken on a flamboyant parade.

Over the first nine days, pilgrims visit nine stations where they take a sacred bath, either very early or at night. These stations are placed at the confluence of rivers, near temples built on the nearby shores. On the eighth day, animals are sacrificed from sunset to dawn: rams, goats, buffalos, roosters and ducks are decapitated. Blood flows freely and is sprinkled on all that's useful, including all kinds of vehicles, even airplanes.

At the beginning of Dashain, the world is empty and time stands still. After 10 days, the society is reborn, the demons have been expelled and the king consecrated anew. Dashain is over at the full moon and life resumes at its usual pace.

Nepalese are fond of esoteric practices and Dashain is replete with them: the whole universe is in the hands of opposing energies: one masculine, the other feminine (as in the yin and yang, heaven and earth, or Shiva–Shakti dualities). The universe is seen as kept in balance by these opposing, yet balancing, forces. Behind the feasts and the games lies an elaborate symbolic system.

The great swing, sometimes very simple, elsewhere richly decorated, is viewed by the naive tourist uninitiated to the world of symbols as a rustic toy or perhaps as a work of art. However, that object, which seems a mere child's toy, is also the symbol of the balance between the two poles of opposing energies. A sense of common identity emerges in the community from participation in the festivities.

As we shall see, the yeti is the focus of a celebration as important as it is laden with symbolic resonance: the Dumje festival.

7. In the Footsteps of a Myth

Documentary film-maker Véra Frossard had been living in Katmandu for six years when she wrote a short but highly revealing book: *The Memory of the Yeti: On the Footsteps of a Myth*.[1] It is a pity that this book should be so poorly known; its simple and precise presentation and its avoidance of embellishments make it a remarkable work.

Véra Frossard relates how she came to enquire about the yeti, following a meeting with a French resident of Nepal who had been living there for 35 years. According to him, the "animal" was only rarely seen. It could climb up to the top of the glaciers. Its cry was a sharp and short whistle that very few people had heard. It mumbled while eating and exuded a fetid smell. The yeti walked on all fours, like a chimpanzee, except in the snow, to keep its fingers from the cold. Its legs were bowed and, when standing erect, it reached 1.60 meters (5 feet 2 inches). Its arms were extremely strong. Finally, her informer said that the yeti was to be seen only in the five or six days preceding the full moon.

Frossard was so puzzled by this description that she decided to enquire for herself, camera in hand. On her low budget she could afford only a single guide, but was well prepared for the quest, used to the altitude, physically strong, and fluent in Nepalese. Newcomers suffer from altitude sickness from 3000 meters (11,000 feet) up, while those having lived in the mountains can climb further without having to rest and acclimatize themselves. At 4500 meters (17,000 feet) there is snow even in the summer and oxygen levels are low. The body needs time to increase its number of red blood cells to ensure enough oxygen is taken in. Sherpas are well acclimatized to high altitudes; presumably, so is the yeti.

Book cover of Vera Frossard's *Memory of the Yeti*.
PHOTO: Editions L'Harmattan

Like another great trekker, Alexandra David-Néel, whose books she greatly admired, Véra Frossard could endure difficult, eight- to 10-hour hikes. She spoke to many people, for example, to a 56-year-old lama who told her, in spite of his reticence, that the yeti was really a god. Reports generally agreed: the yeti were much more numerous before the arrival of tourists in the 1970s. They sometimes came down into the fields, messing up the crops, strangling yaks and eating parts of them.

Frossard was lucky to meet a middle-aged woman, Lapka, who had had a close encounter with a yeti when she was 19. She was herding her yaks and had just put potatoes to cook on the fire. Feeling a presence, Lapka turned around; the yeti grabbed her by the neck and threw her into the river. The water was cold—8°C (46°F)—and she lost consciousness. When she came to, she decided to remain hidden in the water. As night fell, she crawled back to the other shore, where she played dead. But, she said:

> The yeti had killed all my yaks, a three-year old, a *nak* (female) and three other. He had torn open the chest of the three-year old and was sucking its blood although not eating its meat. It wasn't very tall, about 1.50 meters (~ 4.8 feet) but very wide. Its skull was pointed and the short hair on each side of its forehead was clearly parted in the middle. Its smell was very strong, even from where I was observing it. Its body hair was dark, except at its waist were it seemed worn out, and on its chest were it was much lighter.[2]

It took Lapka a while to recover from this nightmare—the yeti did not leave until dawn. At her mother's house, a lama came to heal and exorcise her. Lamas study theology, occult sciences, sacred texts, astronomy and medicine. Many observers have commented on their remarkable psychic abilities. Véra Frossard herself noted that the monks are capable of remarkable intellectual achievements, such as extensive debating contests; on the other hand, while in the physical domain, "they quickly reach through a series of short hops a trance that can last for hours, during which they cannot be interrupted or spoken to."[3]

According to Frossard, such achievements are merely a means to reach the monks' goal. She also speaks of the *toumo*, a practice that allows one to maintain an inner fire. The explorer Alexandra David-Néel had been initiated to this practice during five winter months in the 1920s, at an altitude of 3900 meters (14,000 ft), dressed in a novice's thin cotton dress (*rés kyang*)—from which came the expression *toumo reskiang*. Alexandra David-Néel described the "masters of the art of toumo, sitting on the snow, night after night, naked, motionless, lost in meditation, while the blizzards of winter howled around them."[4] David-Néel's toumo apprenticeship saved her life and that of her traveling companion, her adoptive son the young Tibetan lama, Yongden. (See also, chapter 18)

It is also thanks to that inner fire that paleontologist Stella Swift, the heroine of Philip Kerr's novel *Esau*, was saved from freezing to death in the foothills of Annapurna. An Indian holy man, dressed only in a frayed tunic, warmed her up with his hands:

> The power of his hands seems to draw from a deep internal source, so strong that it appeared to be the life-force itself.[5]

The study of so-called traditional societies shows that each community has a network of basic tenets, which guide their whole life and does so quite effectively. This is clear from medical, psychic and spiritual perspectives. It would be unwise to mock the bizarre habits of remote people. I recall conversations with a neighbor, a Catholic priest, who told me about really bizarre exorcism ceremonies practiced near Nantes hardly 25 years ago, practices confirmed by another friend, a physician from the same area of France.

Lapka, the young woman affected by her encounter with a ferocious yeti, also benefited from a session of exorcism by a lama, but only with temporary effects: questioned a few years later about the event, she suffered a week-long relapse that required new treatments.

Véra Frossard put together a 52-minute film of sequences shot in Nepal. Her document is well worth seeing. She speaks, for example, of westerners' "King Kong syndrome," adapting data about the yeti to their own ideas rather than to those of the Sherpas who have lived in its proximity for centuries. Sherpas think of the yeti as a god, which is probably why they are not listened to. One should, on the contrary, enter into their worldview (scenery, people, fauna and flora), some aspects of which are millennia old.

Lapka reiterated that a god should not behave like that, a belief typical of the Sherpa: as a consequence of her making the yeti a god, scientists did not take her seriously. However, if taken from a different point of view, her testimony offers a path to the imagination of these people and how they construct a myth.

1 Editions de l'Harmattan, 2004.
2 Véra Frossard, op. cit., p.40.
3 Véra Frossard, op. cit., p.43. Such an ability to run quickly in small hops is reminiscent of the yeti!
4 Alexandra David-Néel, *Voyage d'une Parisienne à Lhassa*, p.164.
5 Philip Kerr, *Esau*, London: Chatto & Windus, 1996, p. 334.

8. The Dumje Festival

For a start, the people of Kumjung hang multi-colored flags to the branches of their garden. The next morning they bring joyful banners to the monastery, where the long horns sound their bizarre tones. There is a crowd of people of all ages in the monastery yard. This is the feast of Kumbila, the god of the mountain. Masked men dance to music played by the lamas. At the sound of the gong, all the spectators throw rice. Each family offers a silk scarf to Kumbila who, after his final dance, re-enters the monastery, ending the first day of the festival.

The holiday lasts four, seven, even nine days, depending on the village. The exact date varies according to the lunar calendar, but the key time is the night of the full moon after the summer solstice. On that day, work in the fields ceases; the crops, and with them the future of the farmers, is left to the will of the gods. On this profoundly Sherpa holiday, solemn rites of exorcism are performed to ward off evil and natural catastrophes.

These rites hark back to early shamanic days. The usual practices, such as the construction of a "devil trap," built from threads crisscrossing on a wooden frame, or an offering to calm a slighted spirit, no longer suffice. There is a need for elaborate festivities, uniting the energies of the whole community as a powerful shield against misfortune. It is interesting to note that the Sherpa devil trap has its analogue in the dreamcatcher of North-American natives.[1]

Each year, a number of families—more precisely eight from each village—are responsible for financing the Dumje festival. This is a heavy charge, but each family is proud to play an important role for the benefit of all. This gift to the community is similar in a way to the potlatch of the Pacific Northwest.

During Dumje, the dances, the women's festive apparel, the men in their Sunday best sporting cowboy hats, and of course the beer (*chang*), all contribute to the merriment. Some rites are performed by monks within the monastery walls and are not open to the people. They belong to the esoteric, or secret, world. Other rites, exoteric, are open to everyone's participation. On the third day of Dumje, the man who dresses as a yeti makes his appearance. Frossard, the filmmaker, visited him at home; he is Tibetan, his wife Nepalese.

> I am the one who has carried the yeti's skull in Kumjung for the past fourteen years, one day each year. The people of the village picked me and they pay me with rice, beer or money; nobody really wants to impersonate the yeti.
> "Why?" asked Frossard.
> Because they are afraid...[2]

During the afternoon, there are more dances in the monastery yard. Two of the dancers are dressed as skeletons. But it's really the following day that the yeti-man makes his grand entrance, wearing a sheepskin coat, a black mask and, on the top of his head, the yeti's skull, which looks just like that drawn by Hergé in *Tintin in Tibet*. That skull is a replica; the "real" one is kept in a metal locker in the monastery.

The villagers at first hesitate to approach the yeti. They soon gather courage and expel from the monastery the fearsome destroyer of crops. They then also leave the monastery yard and a procession of monks, musicians and villagers pursue the yeti to the limits of the village.

Back in France a few months later, Véra Frossard discovered a similar celebration when she attended the Festival of the Bear, a relic of a ritual formerly common in all Europe. The festival is held each year in February in Prats-de-Mollo (Pyrénées-Orientales), France. A man disguised as a bear terrorizes the village. He is caught and the celebration ends with the shaving of the bear: its animal nature is removed so as to bring it back to the civilized realm.[3]

Closing her story, Véra Frossard reflected on the words of an old Buddhist monk: "There is a yeti deep in every man's spirit. Only the wise are not haunted by it."[4]

A color supplement complements her book, with pictures of

scenery, houses, people, costumes and relics. The author overcame cold and fatigue to reach people involved in the quest for the yeti, for example, the Sherpa Gyalzen, who acted as a guide for researcher Robert Hutchison. Gyalzen distinguished two types of yetis: a short one, the *mitey*, and a tall one, *chuty*, speaking of the wild man in general as *migo*.

Testimonies collected by Véra Frossard suggest that the kidnapping of women by male yetis is just as frequent as the capture of men by female yetis. There are sometimes offspring, showing that the two species are genetically compatible.

Chinese Tarzan, the monster abducts the heroine. PHOTO: Chinese comic book, 1986

It is only because she knew how to approach people that Frossard managed to acquire such information. Speaking the language made a big difference. She probably managed to explain—perhaps even to show through videos—what she wanted to communicate to the public, creating a functional "ethno-dialogue" if not exactly reaching a shared anthropology. She especially succeeded in communicating the symbolic role played by the yeti, as traditionally represented by a masked villager. It becomes quite clear that the authenticity of the skull used during the Dumje celebration is of little relevance. What matters is the resemblance, which suffices to identify the amateur actor with the yeti. At that point, the yeti stands for more than just an animal. It is the symbol of the threat of bestiality, that tendency which leads from beer to orgy. The beast, or rather its brutality, must be expelled from the village through dances, music and rituals. The goal is to bring together the animal and rational natures of man, its primeval urges and civilization.

However, the spread of modern consumer society habits has had its impact on traditional customs. Gyalzen notes that the yeti is more rarely seen today and that young people no longer believe in its existence. The actual physical presence of the yeti would seem to be the key to the continuation of the Dumje rituals.

In any case, René de Milleville, a Frenchman who lived in Nepal for 35 years and wrote extensively about it, remained convinced that the yeti would some day be identified. In 1985, he provided Prof. Michel Tranier, of the mammology laboratory of the Paris Muséum d'Histoire Naturelle, a handful of reddish hair. Tranier concluded that they belonged to a primate closely related to the orangutan.

What are we to think of the ritual behavior of the yeti-man? Might it be the expression of an antagonism between, on the one hand, an ancient shamanistic attitude seeking violent power through magic and sorcery, or the ascetic and pacific nature of Buddhism, favoring detachment from the world?

The man who wears the animal skin, the pointed skull and the yeti mask is exposed to the influence of dark forces. He takes upon himself the demonic powers, which the community seeks to expel during Dumje. The danger that he may himself fall victim to these powers is such that there are few volunteers for the role. During Dumje, many spectators avoid approaching or even looking at the yeti. It is the powers that it represents rather than its actual resemblance to the animal that really matters.

---

1 Véra Frossard also speaks of the "delicious Tibetan bread." Similarly, North American Natives fry an unleavened bread, bannock. These simple foods acquire a symbolic meaning as links with the nurturing Mother Earth.
2 Véra Frossard, op. cit., p.52.
3 For further details, see my book *Sasquatch/Bigfoot and the Mystery of the Wild Man*, Part IV, Sasquatch and the World of Mythology.
4 Véra Frossard, op. cit., p. 95.

9. Esau the Hirsute

As a preamble to scientific speculations about the nature of the yeti, let's take a look at Philip Kerr's novel *Esau*. (1996). Kerr's main protagonist, star alpinist Jack Furness has conquered Everest without bottled oxygen. On a subsequent expedition, he falls into a crevice. His climbing mate is killed, but Jack survives and returns with a nearly intact skull found at the bottom of the deadly pit. He hands the fossil over to his friend Stella Swift, a paleoanthropologist at the University of California at Berkeley. At first glance, she is reminded of *Paranthropus robustus* or perhaps even *Gigantopithecus*.

Paranthropus, as the name suggests, was similar to man: *para* = close, *anthropos* = man.

Prehistorian Leroi-Gourhan suggests, "the evolution of the anthropic skull seems to reflect a triple process: the freeing of the back part of the skull arising from the adoption of an erect posture; an expansion of the forehead following the progressive shrinking of the dentition; an increase in the volume of the brain up to the Neanderthals, followed by an expansion of the frontal lobes without increase in volume."[1]

Robustus was distinguished by a prognathous appearance, with large jaws and a prominent sagittal crest. Although mainly a herbivore, its powerful jaws—it has been nicknamed the nutcracker—allowed it to chew tough roots as well as insects, rodents and small game, basically eating whatever was available. *Robustus* was of modest stature, less than 1.50 meters (4 feet 10 inches) tall and weighing between 25 and 45 kilograms (56 to 100 pounds); it used simple tools but did not master the use of fire.

Robustus lived between 2.2 million and 1.0 million years ago, a forerunner but not a member of the *Homo* family. Other species were evolving at the same time: *Australopithecus garhi*, soon to disappear; *Homo habilis*; *Homo ergaster; Homo rudolfensis;* and even *Homo erectus*, which would survive long after *robustus*.

Should *robustus* be distinguished from its predecessors, the Australopithecines, simply because of its large jaw? This is a matter for discussion among paleontologists. There is no specific reason known for the disappearance of *robustus*. It certainly possessed a number of useful traits: a varied diet, walking erect while still able to climb trees, even mastery of fire for *Paranthropus boisei*, a variant discovered in 1959 by Mary and Louis Leakey in Tanzania.

Giganthopithecus, a giant ape, which lived 400,000 years ago, weighed up to 600 kilograms (1450 pounds) and measured up to three meters (10 feet)! A bamboo eater, like the giant panda, it may have been hunted by its cousin *Homo erectus*. Its remains—three jawbones and a million teeth—have been found in China and Vietnam. Competition for food with the panda and *Homo erectus,* as well as mortality at the hands of *Homo erectus* hunters, may have been responsible for the demise of *Giganthopithecus*. Below a certain population level, the species would no longer have been able to survive. However, *Gigantopithecus* might have survived in the Himalayas (as the yeti) and across the Bering Strait (as the sasquatch).

Book cover of Philip Kerr's book *Esau*. PHOTO: Chatto & Windus, 1996

Isolation would have protected these hairy giants; otherwise *Homo sapiens* would have wiped them out, as it probably did Neanderthal man. However, it would be a mistake to imagine that only the genus *Homo* is capable of systematic destruction. The author of the novel *Esau,* of which Jack Furness and Stella Swift are the heroes, quotes Jane Goodall, who studied the warlike nature of a group of chimpanzees systematically eliminating the members of an enemy group. Jane Goodall thought that the process of extermination would have been much more rapid had the chimpanzees benefited from human weapons.

In the novel *Esau*, researcher Stella Swift consults a computer specialist to obtain a digital image of the brain that used to fill the fossilized skull discovered by Jack Furness. The computer model makes it possible to view the various parts of the brain, to examine details, and to view the various lobes from different angles. Stella is especially interested in an area dear to her heart: Broca's area, where language ability is thought to reside in humans. That same area is clearly present in the skull under examination.

Following linguist Noam Chomsky, Stella believes that man possesses a boundless capacity of expression through language; his powers of thought and imagination are without limits. A child not only repeats what he heard; he also invents new sentences. Stella firmly believes that the emergence of consciousness is linked to the acquisition of language. Her ultimate ambition is to discover a fossil demonstrating the antiquity of linguistic abilities, and consequently of human consciousness, in an era which she thinks of as "The Dawn of Humanity." Stella's plans for an expedition in Tibet are clearly not based on the lure of profit or publicity.

The expedition is quickly put together; the ease with which it is financed is nothing short of miraculous. Stella does not suspect the covert role of the CIA in assembling the scientific team, one of whose members is a secret agent charged with finding out about Chinese activities in Tibet. Have the Chinese, following their occupation of Tibet in 1950, perhaps built secret factories, missile bases and radar stations? Might they perhaps want to take advantage of the India–Pakistan antagonism to invade Nepal?

Spying is a common theme in today's popular novels. The presence of an evil CIA agent offers the prospect of murders and mindless cruelty, regretfully too often overemphasized.

Unfortunately, scientific research is often subject to world events. Some areas become inaccessible due to armed conflicts. The fauna, and even the flora, may be affected. Catastrophes, natural as well as artificial, such as the 1986 Chernobyl disaster, set difficult conditions. Thus, vast areas, even entire provinces, may become a no-man's land. Travelers cannot obtain visas; many stay away concerned about safety. Whole regions return to the exclusive domain of a few natives and especially of animals. The rare visitors may again glimpse elusive animals, such as the snow leopard, in the Himalayas.

One of these large cats is shot down with a hypodermic dart by

Stella's team. The encounter with the leopard is a prelude to that expected by the reader: a meeting with the yeti, here called *Homo vertex*, Man of the Summits. In an atomic conflict, *Homo vertex* would survive a nuclear winter, already adapted as it is to live in semi-arctic conditions. It would then replace *Homo sapiens*...

In the novel, a feeling of panic comes over the expedition as the leader of the Sherpas has a close encounter—less than 30 meters (100 feet) away—with two yetis, large corpulent males, at least two meters tall, resembling gorillas with red-brown hair. Their heads are large and pointed, their faces hairless, flatter than a human's but not as flat as a gorilla's. These yetis stink terribly, a condition worsened by their coprophagic habits. The Sherpa notes that he has witnessed a yeti eating its own excrement. One of the scientists remarks that some great apes also do this in order not to waste any nutrients.

Soon afterwards, the explorers discover another feature of the yeti. They see a female smash a marmot against a tree and then eat the intestines before throwing the rest of the carcass away. Yetis are carnivores, as are also chimpanzees on occasion. The story is peppered with zoological details, which add somewhat to its plausibility, as well as information about the Sherpas' religious habits. An image of the green goddess Tara is supposed to protect the Sherpa leader. Tara, the highest form of the eternal feminine, is the equivalent of the Hindu Shiva.

Not surprisingly, the realism of the adventure evaporates when we discover the yetis' refuge: a small valley, difficult of access, in the crater of an extinct volcano, with fertile soil and rich vegetation—a magnificent mountain oasis that the Sherpas name "The Sacred Forest." A wise *sadhu* reigns peacefully over the yetis, viewing these creatures as so many portals through which he can contemplate the marvels of the universe.

The heroine, Stella Swift, is brought back to the scientists' base camp in a rather unexpected fashion: in the arms of a yeti, like Fay Wray in those of King Kong. She returns from a great height at breakneck speed as the yeti slides down the snowy slope on its rear end, as on a giant toboggan—a spirited fantasy heralding a happy ending to the novel.

Philip Kerr's novel takes the reader to the core of some of the twentieth century's anthropological concerns, those which led Charles Dawson, Raymond Dart, the Leaky family, and more recently George Nutall and Jared Diamond, on the path of revealing enquiries. We recall that Charles Dawson was the one who discovered, in 1912, the remains of the Piltdown Man, a fabrication made up of a human skull and the jaw of an orangutan. It was only in 1955 that the hoax was revealed and it took some time for paleoanthropology to recover its reputation. The shadow of Piltdown Man was still hovering over the field when Raymond Dart discovered, in 1924 at Taung in South Africa, the fossilized remains of the first *Australopithecus*, to which he gave the specific name *africanus*.

Louis Leakey (1903–1972) was the patriarch of a clan of paleontologists and paleoanthropologists (five in all). His granddaughter Louise (born in 1972) continues the work begun by her grandparents in eastern Africa 70 years ago. Their discoveries have opened up new perspectives in human prehistory. Their family history is a modern scientific and sociological epic.

Bacteriologist George Nutall was the one who first demonstrated that human and chimpanzee antigens were nearly identical.

Jared Diamond, the author of *The Third Chimpanzee* (1991), teaches physiology at the University of California. In his view, the future of mankind—the third chimpanzee—is threatened by the dangers of genocide and ecological holocaust.

---

1   Quoted from *Le Grand Robert de la Langue Française,* tome 1, p. 571.

10. Messner's Yeti

We now pass from fiction to a factual account by Reinhold Messner.

> Suddenly, as silent as a shadow, a silhouette emerged, about ten meters in front of me, in the middle of a thicket of rhododendrons. 'A yak,' I thought, immediately relieved at the idea of meeting with Tibetans, and sharing a hot meal and finding a warm place to sleep. But the creature remained immobile. Then, silently, it began to run through the forest, now disappearing, then reappearing, ever more swiftly. Neither branches nor ditches slowed it down. It was NOT a yak![1]

Biplane over Everest. PHOTO: Author's file

Thus did Reinhold Messner describe his encounter with a mysterious creature. Between 1970 and 1987, Messner conquered the highest and most challenging mountains of the world. Together with a partner, he climbed Everest (8850 meters [29,000 feet]) without oxygen in May 1978. He repeated the feat alone in 1980, again without oxygen, carrying only a light backpack. He continued his exploits, climbing all of the world's peaks above 8000 meters (26,000 feet), demonstrating again his extraordinary adaptation to the summits.

Once all the summits were reached, Messner turned to exploration, often alone. He traveled through eastern China, Tibet, Nepal and Buthan. In 1993, he crossed Greenland (2000 kilometers [1250 miles]). In 2004, he trekked through the Gobi desert in Mongolia, another 2000-kilometer hike.

Like Edmund Hillary before him, Messner sought to challenge both himself and nature. "It is only nature which I must respect, the nature that's within me as well as that external."[2] Different times meant different methods: Hillary's expedition included 350 porters and extensive supplies, including oxygen (until the 1970s, Everest climbers carried 50 kilograms of oxygen tanks each, to be used above 7000 meters [22,965 feet]); Messner's support team, 27 years later, included a liaison officer and an interpreter, both Chinese. He was one of the first to obtain permission to climb the north face of Everest. In addition, an American woman who worked in school construction for the Sir Edmund Hillary Foundation, accompanied him as first aid/cook/general helper.

Hillary, we recall, had developed a close friendship with the Sherpas. He devoted time, money and energy to improve their lives, building 27 schools, two hospitals, 12 clinics and a few bridges. Sadly, these accomplishments were shadowed by the death of Hillary's wife, Louise, and their youngest daughter, Belinda, in 1975when their plane crashed after leaving Katmandu.

Messner also suffered a serious loss: in 1970, after a successful ascent of Nanga Parbat (8126 meters [26,660 feet]), his brother Günther disappeared during the descent. A few days later, Reinhold had to have seven frozen toes amputated. Günther's remains were found only in 2005 and were incinerated by Reinhold on the Nanga Parbat. In his novel *Esau*, Kerr drew liberally on Messner's adventures as an alpinist.

Hillary and Messner, conquerors of the Everest, shared similar fates. Besides his exploits as an alpinist, Messner also explored vast regions from Sichuan to the Pamir, from Bhutan to southern Siberia. Often traveling alone from 1985 to 1997, his long treks aimed at tracking the "abominable snowman," also Hillary's quest.

Messner's investigations led to his book *My Quest for the Yeti*, (1998). In that year, the author handed all his documentation to

American mammologist George Schaller. Messner now faced new dangers: the irony of the press and the hostility of the elite of mountain climbers.

We now return to that late afternoon in July 1986 when Messner glimpsed that fugitive silhouette. He then found an enormous footprint in the dark soil. He took pictures, recalling similar photos taken by Eric Shipton in 1951. That same evening he found four more prints. Pushing through the juniper bushes, he heard a whistling noise:

> From the corner of my eye, I saw a standing silhouette, between the trees on the edge of the clearing, in an area where short bushes covered the steep slope. The creature moved fast, leaning forwards, noiselessly, disappearing behind a tree to reappear quickly in the moonlight. It paused and looked at me. I heard that whistle again...a kind of angry whistle. For a brief moment, I saw its eyes and its teeth. The beast was standing, menacing...It was covered with hair and had short legs; its powerful arms hung alongside its body down to its knees. I estimated it to be more than two metres tall and it seemed to weigh much more than a man of similar stature. I was surprised to see this enormous mass move with such agility. It was now moving towards the edge of the escarpment. I felt relieved. I was mainly astounded. No human being could have been running in this manner in the night.[3]

Such is the beginning of Messner's account, as he climbed a valley of eastern Tibet, carrying only a sleeping bag, a flashlight, a penknife, a safety blanket and a camera. He was looking for a village to spend the night. Without a tent, he would have to sleep under a tree or in a cave if no one took him in. As he approached a village, Messner discovered the truth of the Tibetan proverb, *the guard dog is one with the door as a woman with her jewels.*

Hounded by the Tibetan mastiffs guarding the village, he took refuge in a loft from which the villagers expelled him. After a few words, they welcomed him and offered him *tsha* (tea), *tara* (butter) and *tsampa* (grilled barley malt). In answer to their questions, he described his encounter with a creature as large as a yak. His Tibetan hosts exclaimed in frightened unison, "*Chemo?*"

Back on the road, a one-eyed villager guided Messner. For him, the *chemo* was a nocturnal creature that looked like a bear or a huge ape. Like the villagers, it ate barley, meat, fruits, berries, vegetables and nettles.

From then on, Messner's search took a new turn. The Sherpas, expelled from their Tibetan homeland had brought their stories of the yeti to Nepal. However, he had just encountered a creature named *chemo*. Was it the same as the yeti? Messner began to imagine that the discovery of the yeti could explain the origin of Sherpa legends.

---

1   Reinhold Messner, *My Quest for the Yeti*, p.4.
2   Messner, as quoted by Caroline Alexander, *National Geographic,* November 2006, p. 34.
3   Reinhold Messner, op. cit. p. 7.

11. A Casual Meeting

A casual meeting can be transformative; it can change everything, including the scenery! The yeti no longer shows up in the mountains and high valleys of Nepal and Solo Khumbu. In Nepal, the yeti is now only a ghost. But in eastern Tibet, it is still real.[1]

How can one truly fathom the impact of scenery? It might bring to mind memories of childhood, of love and adventure. It might hide a secret spring, caves, ancient ruins. It is the humus from which rises the rare flower, the refuge of the last few survivors of a hidden species. The region is either graced or haunted by the presence of a being seen as either admirable or evil, sometimes both at once. Are there two faces to the yeti? On the one hand, the killer of yaks and rapist of women, on the other, a timid and inoffensive giant? Seen from that latter side, the yeti would only venture into forested areas in search of salt, finding there the salty mosses required by its metabolism. It would otherwise avoid contact with humans.

Gerald Russell's analysis of yeti excrement suggests that it is not exclusively vegetarian. Bernard Heuvelmans adds, "The evidence shows that the Snow Man clearly has an omnivorous diet: roots, fruits, lizards, birds, small rodents and, occasionally, larger preys—anything that might be at hand in such a dry area."[2]

Heuvelmans also points out the imprecision of the terms used to describe the snowman. Should one detect there the influence of Tibetan Buddhism, according to which the souls of the departed are reincarnated in the bodies of near-humans and should not be addressed by their real names? Besides *mi-gö* (Tibet, Nepal) or *mighu*

(Bhutan), meaning wild man, one also finds *dre-mo*, which means female demon.

Did Messner go on a quest for a real or for a mythical being? He raised the question himself at the end of the first chapter of his book. Over the dozen years that followed his July 1986 encounter he launched a series of expeditions. The search took place on a background of political and military upheavals affecting the entire population of Tibet, as well as that of neighboring countries (Nepal, Bhutan and India) that took in Tibetan refugees. In Katmandu, Messner conversed with Tarchen, his Tibetan interpreter, who had left the country in 1950. Tarchen, a former Tibetan resistance fighter, related how he and some companions had discovered the corpse of a yeti, shot down by the Chinese. The creature had been skinned; its body was similar to a man's, but thicker and more muscular. Messner commented, "Tibet has always been rich in myths and legends, but poor in freedoms."[3]

Messner often refers to the importance of a myth celebrated during the annual Mani Rimdu feast meaning "all will go well." The feast lasts at least three days in Sherpa country; it is an invitation to celebrate the gods of nature. The monks wear masks of the gods. The spectators make offerings of maize to the monastery. In Chiwong Gompa, prayers are offered to Guru Rimpoche, the founder of Tibetan Buddhism. The feast begins with the sounding of the horns. The second day, costumed monks dance a mime of the struggle against the forces of evil, accompanied by an orchestra of horns, conks, flutes and cymbals. On the third day, bread-dough figurines (*tormas*) are burned: the evil forces have been eliminated!

According to Messner, these practices are vestiges of the ancient shamanic religion *Bon*, which preceded Tibetan Buddhism. The blood and the scalps of yetis were mixed with the blood of a horse, a dog, a goat, a raven, even human blood. The blood had to come from the blood of a yeti killed by arrows.

Messner witnessed this ritual at the Khumjung monastery, at the foot of Everest. A man wearing a conical scalp and a sheepskin carries a bow and arrows. It is said that an evil fate may fall upon this disguised man. Messner relates these rituals with great objectivity, and describes them without irony or patronizing. He notes that following the introduction of a deeply esoteric form of Tibetan Buddhism in the eighth century, many elements of the old religion survived—hence this mixture of spiritual and shamanic elements that lies at the root of Tibetan culture. The term esoteric, as used by Messner, describes a form of knowledge that remains, without prior initiation, hidden and inaccessible.

Following 18 months of investigation, Messner concluded that two questions needed clarification. First of all, why had the legend of the yeti spread so widely and durably over such a wide area, from the Pamir to the eastern Himalayas? Secondly, could it be shown that there actually exists a living creature behind the legend? Messner discarded the option of a link to some hominid creature. He was convinced that the chemo was the living proof of the legend.

The shaman of the Grotte des Trois Frères.

---

1   Reinhold Messner, *My Quest for the Yeti,* p. 27.
2   Bernard Heuvelmans, *On the Track of Unknown animals,* p. 193.
3   Reinhold Messner. loc.cit., p. 92.

12. Discoveries

In the spring of 1991, Messner crisscrossed Bhutan with a German TV reporter, a cameraman and a photographer. For weeks they trekked through the high meadows lying between impenetrable forests and icy summits—the ideal yeti habitat.

On a cool and sunny day in May, the team paused at Gangtey Gompa, a small monastery. Crossing the prayer halls during his visit to the gompa, Messner noted the absence of any image of the yeti. He asked to see the tantric room; the monks gave their permission. In the dim light, the team discovered walls covered with fabrics painted with masks and skulls. Below, hung hunting trophies: the heads of elks, bears, boars, sheep, deer—and above it all, the pelt of a "red yeti." This *mygio*, explained the lama, reigns over all the other animals. It was the skin of a young yeti, cursed before removing its heart and skinning it.

Scrutinizing the pelt, Messner quickly concluded that it was a fake, no more than a puppet. "I no longer knew what to think. Only one thing was for sure: the yeti had no place in the concrete, rational world and could only have arisen in such a place. It was either the product of the imagination, or a symbol for a particularly rare animal."[1]

Messner was very disappointed, especially in the light of the sarcasms with which expert "yeti-ologists" greeted each one of his forays. Nevertheless, he did not give up, spending the whole winter of 1992–93 in Nepal. During a visit to the hidden kingdom of Mustang he heard a story about the king having once killed a yeti that was carrying away a young yak. Not much of a reward for all Messner's efforts.

During the summer of 1996, Messner and a few friends traveled to Chengdu in Sichuan, en route for Lhasa. They were struck by the degree of deforestation and by the number of yak corpses seen along the road. They pitched their tent for a few days on the shore of the "divine lake" Yulung Lhantso. This is where Messner overheard the word chemo repeatedly mentioned in a conversation between the porters and members of a caravan.

A man called Lopsang who lived at the edge of the forest told Messner that he had recently seen a chemo. He led Messner along forest paths, across clearings and undergrowth, to examine tracks in the snow, similar but much larger than those left by human beings. The chemo was said to prey on livestock and steal goats, sheep and yaks, but also to eat berries, marmots and even ants.

Messner then paid a visit to a hermit who lived near Lopsang's tent. "He finally stood up and led me to his tiny meditation room, where there were a number of objects: a skull set in silver, a flute carved in a human bone, a drum covered with human skin. Back outside, he told me that a few days earlier he had seen a mother chemo playing with her child and that he had talked to her."[2]

Everyone will recognize there the instruments of an ancient ritual: skull, flute and drum are all implements to assist in meditation and trance. As to the dialogue with a chemo, it is clearly the prerogative of shamans capable of conversing with animals and spirits.

Back at camp, Messner haggled with a woman over the price of a so-called chemo paw that she was trying to sell him. It was in poor condition and Messner recognized it as the paw of a bear. "Under the thin hair, the smoke-dried paw looked vaguely like a human hand: real or fake, the object was the very embodiment of the yeti myth."[3] Messner realized that the nomads took, in all good faith, this moth-eaten paw as that of a yeti and not that of a bear.

Pursuing their trek along icy, wind-blown trails, between 4000 and 5000 meters (13,000–17,000 feet), the explorers met whole families of pilgrims from Tibet or neighboring Chinese provinces (Sichuan, Qinghai, Sinkiang) en route to Lhasa, camping along the road, warming up with hot tea. They burned aromatic herbs and prayed continuously. Prayers become most important during processions to sacred places.

The rapport between men and nature—animals, scenery, mountains—forms the normal spiritual background of the Himalayan

China and surrounding countries.

realm. For westerners, this particular perspective is likely to play a secondary role, a fact of which Messner is well aware and never forgets. He has long understood the links between men and, for example, the ibex or the musk deer, or the *takin*, a bovine resembling the African gnu, which lives between 2000 and 4500 meters (7000–14,000 feet).

Among the familiar local fauna, there lives the Tibetan wild yak, quite rare today, weighing up to a ton and living high in the mountains, from 3200 to 5400 meters (10,000–17,000 feet) in temperatures as low as –40°C. Its domestic counterpart forms the economic basis of numerous Tibetan communities. These animals, long established in the area, are viewed from two perspectives, one being utilitarian, the other symbolic, or even mythical. But these are living, not bloodless myths, still useful, functional and certainly not inert and dusty objects.

Messner is well aware of the constant exchange between the material plane and the mythical domain. He thinks he is close to finding the key to the mystery when a pilgrim leads him to a village hut housing a store. Hanging on the wall, there hangs a stinky, greasy pelt, which Messner recognizes as that of a rare bear living between 3600 and 5500 meters (12,000–17,000 feet)—the chemo. "In other words, the yeti is a chemo having taken on mythical proportions."[4]

1   Reinhold Messner, *My Quest for the Yeti*, p.101.
2   Messner, op. cit., p. 130.
3   Messner, op. cit., p. 131.
4   Messner, op. cit., p 132.

13. Messner Perseveres

In April 1997, Messner once more left Chengdu, in Sichuan, for Tibet. Accompanied by a guide-interpreter, he was confident that this expedition would allow him at last to conclude his quest. Reaching a broad valley, they decided to pitch their tent. The green meadows of Dora spread before their eyes. One day, Messner and his guide spied with their binoculars a young chemo at the edge of the forest. They named him "White Head."

The people of that region describe the chemo in the same way as Tibetans speak of the yeti.

> It is a nocturnal being, usually solitary; its pelt is sometimes dark, sometimes light and changes in color with age. He often walks on its hind legs and often stands erect like a man. Its droppings look like human excrements, although they contain chewed-up bones and the hair of small rodents. It usually lives below the snow line, even in high summer, and only ventures onto the glaciers when strictly necessary. They can be heard and seen near villages, especially in the spring and during the long winter months.[1]

Day after day, trekking through the majestic mountain scenery, Messner observed and inquired. One day in May, he and his guide rode their horses to the Sosar Gompa, home to about a hundred lamas. Above the entry gate, hung two stuffed animals: a yak and a chemo. The chief lama granted Messner permission to photograph the chemo. Taken down from its hook, the chemo looked impressive, even down on the ground. The monks stared at it respectfully. The boldest lama grabbed the mummified chemo and started dancing with it as the others clap their hands. Their song was like the whisper of waves spreading on the sand. One imagines hearing a silent approval of the bold dancer. One should not forget that for many Tibetans seeing the yeti is an evil omen.

Messner finally concluded that the yeti is a brown bear, a creature that the mountain people of Tibet call chemo and which they have always held in high respect. Its survival, like that of many others, is linked to the respect of nature. "Without wilderness, there is no yeti."[2]

Messner always strives to describe nature with great precision. Raised in the bosom of alpine poetry, he is very good at it. He easily depicts the scenes of the daily life of villagers and nomads. In their huts or under canvas, one can see the people living their lives, among their hangings and utensils, gathered around the fire or huddling together for warmth. Even their gestures, their facial expressions and their intonations reach us. The animals, the prairies and the woods feel familiar. He successfully conveys the essence of those Himalayan countries that he loves. Therein lies one of the greatest strengths of his work. Perhaps it is the essential element: that country of the yeti, a unique locale where the creature can flourish—or rather, just survive in this day and age. Its survival appears linked to that of the local population and their lifestyle; realizing, of course, that no lifestyle is completely unchanging.

Each trip through Tibet distanced Messner from the media image of the yeti. Messner suffered the consequences: Wild man specialists began to despise him; sharp tongues insulted him and the media continued to exploit the "abominable snowman" theme. Messner persisted with his conclusion that the flesh-and-bones yeti is a chemo, a rare Tibetan bear.

Let's hear what François de Sarre, editor-in-chief of *Bipedia*, had to say about Messner's proposed thesis that the yeti is a bear. De Sarre

Ursus speloeus (The Cave Bear). 19th century French advertisement for chocolate. PHOTO: Author's file

pointed out the crossover between the real *Ursus arctos* and the yeti, nocturnal demon product of the human imagination.

> Of course, the bear might have inspired a number of Himalayan legends, but it seems improbable that it should be at the bottom of <u>all</u> reports on the yeti.[3]

De Sarre regreted that Messner had focused on a single possibility, that of the brown bear *Ursus arctos isabellinus,* which neophytes could easily confuse with the black bear *Ursus thibetanus*. That being said, even if the bear could have given birth to numerous legends, it could not by itself be the solution sought by so many investigators. A bear could not encompass all the clues gathered over the centuries: tales, rituals, encounters, footprints…Nevertheless, who could dispute Messner's right, on the basis of his extensive experience, to favor a specific hypothesis, however restrictive it might seem.

In any case, François de Sarre appreciated the descriptions of the vast areas traversed by Messner, often on his own. He also noted the author's vivid portrayal of the people of the Himalayas, the mark of an exceptional field observer. From that perspective, de Sarre acknowledged the merits of Messner's book and recommended it to the amateurs of cryptozoology.

However, although the mysteries of the Himalayas might now seem somewhat less opaque and the religious practices linked to the yeti rather less strange, from a scientific perspective the question of the existence of the snowman remains unanswered.

---

1 Reinhold Messner, *My Quest for the Yeti,* p. 132.
2 Reinhold Messner, op. cit., p. 164.
3 François de Sarre, "Is the Yeti the Brown Bear?" *Bipedia,* Mar. 23, 1999, p. 9.

14. A Parenthesis: Links to the Pacific Northwest

I cannot possibly hide the contents of the correspondence that just came in from Seattle, USA. Among the documents included, there is one that expands on Messner's comments about the role of animals. We recall the importance of the ibex, the musk deer, the takin and the yak in a hierarchy dominated by the yeti.

The document that caught my eye told of the making of a blanket, to be worn as a mantle, from the hair of mountain goats. It was the

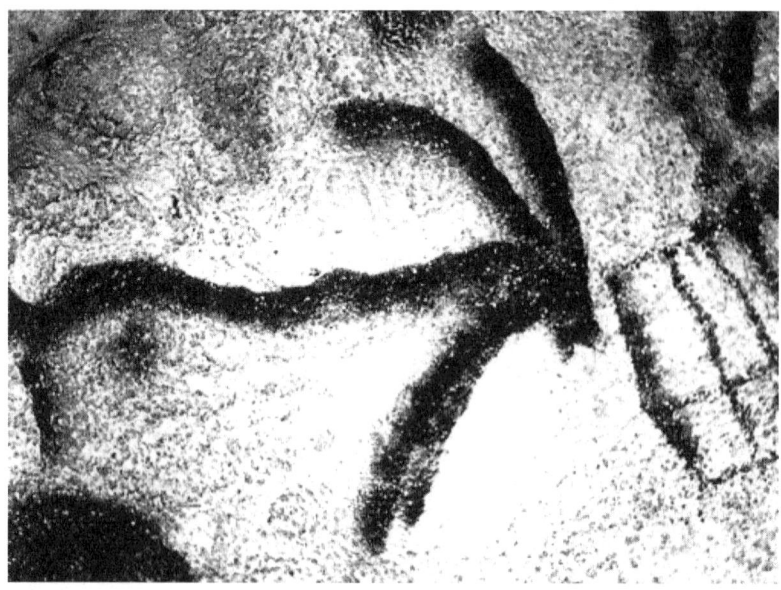

A stylized ibex from the Lascaux cave. Photo: Author

first time in many generations that such a garment, a product of ancestral craftsmanship, had been publicly presented to the members of a Puget Sound tribe of Native Americans. Only a few master weavers had preserved the technique of hand weaving such a blanket.

A presentation ceremony had been staged for this masterpiece, weighing over 15 pounds. It took place on January 27, 2007, in the longhouse of Evergreen State College. It began with a procession, chants and prayers. The blanket was placed successively on the shoulders of all the elders who were honored. They had to remain sitting because of the weight of the garment. One of them expressed a deep feeling for the energy emanating from the blanket. The artifact is rich in meanings; it revives a forgotten art of weaving; it gathers the scattered remnants of the Native community; it also represents the will of the Salish to preserve their traditions. "It is part of our generation's efforts to recover our identity."[1]

Beyond the mountain goat and the mantle we can detect in this new weaving the principles underlying a people's identity. The scenery changes, but the principles remain. The symbolic meanings, here of a mantle, there of a mysterious creature, are universal. Understanding this meaning makes it possible to understand the community, to join it, rather than remaining an ignorant stranger forever.

---

1   Michael Pavel, a member of the Skokomish tribe, as quoted in "Blanket brings sacred change," Lynda V. Mapes, *The Seattle Times*, January 28, 2007. This blanket is now on exhibit at the Seattle Art Museum.

15. Siberia

In 1995, Reinhold Messner undertook an expedition into the Altai range (4500 meters [15,000 feet]) of southern Siberia. This is the region, south of Novosibirsk, which abuts Kazakhstan in the west and Mongolia in the east. As we shall see, Messner's explorations in Russia and China were the logical follow-up on his earlier expeditions in the Himalayas.

In his book *La Lutte pour les Troglodytes*,[1] Professor B.F. Porchnev drew a map showing the migration routes of relic paleanthropes, "ancient men," which he identified with latter-day Neanderthals. Porchnev pointed out the vastness of the area that stretches from Kazakhstan, through the Altai, Lake Baikal and the Khingan range, all the way east to the Verkhoyansk Mountains—an area rich in tales, legends and eyewitness reports of manlike creatures preying on reindeer herds. These beings are said to spend their summers in the extreme northern parts of Siberia, on the shores of the Chukotka River.

Porchnev quotes the archaeologist A.P. Okladnikov:

> The Chuchunyas are a tribe of half-human, half-animal beings which formerly lived here, in the North, and which are still seen occasionally, although rarely. Their head was as if stuck to their body, without a neck. At night, they would suddenly show up at the top of a cliff and throw rocks on the sleeping men and steal a few reindeer from their herd. Makarov, a Yakut hunter, states that he had seen caves inhabited by the Chuchunyas on the right bank of the Lena River...In those natural shelters he had found the horns and skins of reindeer which they had devoured.[2]

This report is typical of the eyewitness accounts emanating from Yakutia. Later, in 1908, the young mineralogist P.L. Dravert left notes about the herculean natives of the Lena River. According to a Yakut source, these hairy creatures ranged as far as the Aleutian Islands, known as "the warm islands."

> Once, in the land of the Chukchi, the sea left on the beach the body of a very hairy man from the warm islands…He lay there for an entire day. A respected shaman was the only one who saw him, on the following night, rise from the beach and walk three times around the huts of the Chukchi before departing.[3]

The hairy man is apparently endowed with exceptional abilities. He is feared, and only a shaman could approach closely enough, at night, to observe him.

Jean Servier's book.

Among the Yakuts, shamanism has become very elaborate. A range of specialties have developed to deal with every spiritual tradition. There are, for example, "white" male shamans and "black" female shamans; others specialize in healing, counseling, dream interpretation, or prophecy.

Upon the appearance of a hairy wild man, it was probably the shaman's responsibility to find out whether the unknown intruder was likely to cause any harm.

The white shaman's role was to relate with spirits and the gods of heaven. This distinction between white and black shamans is also found among the Buriats, somewhat closer to the Altai range explored by Reinhold Messner.

The shaman understands the language of animals, which also allows him to converse with spirits. He must develop friendly relations with helping spirits, which join his own protecting spirit and enhance his power in his struggle against demons.

Are the hairy men, half-man and half-animal, among the shaman's helping spirits? Or might they belong with the demons? The chuchunyas mentioned above seem to be of a rather complex nature, requiring the intervention of an experienced specialist. Such a shaman would know what to do should chuchunyas happen to threaten the souls of the tribe's people. The necessary rituals find their counterparts across the ocean, among North American natives.[4] The similarity of the practices found in such diverse regions is striking:

> Shamanism in these regions (Mongolia and Siberia) is closely linked with the religions and beliefs found in two widely separated parts of the world. North America was probably peopled by Siberian hunters who crossed Bering Strait while it was still a bridge between the continents. The shamanism of American Eskimos is almost identical to that of the Siberian Chukchi. The Mongolian shamanic tradition is close to the pre-Buddhist Tibetan religion bonpo and to various religions surviving in Nepal and southeast Asia.[5]

Ainu man, circa 1880.
PHOTO: Author's file

Before leaving Siberia behind, let's consider for a moment the Ainu, an ancient folk still found today in northern Japan, on the island of Hokkaido. The Japanese parliament formally recognized the Ainu people on January 16, 2008. The Ainu are characterized by their pale skin, their non-slanted eyes and by their hairiness. Ainu men grow abundant beards; after puberty, Ainu women decorate their skin with tattoos. The Japanese used to call them "the hairy savages." They are known to be related to Tibetans and to Mongolian and Siberian natives.

For the Ainu, everything is endowed with spirit, including even the objects they fashion. However for such objects to possess a soul, they must be decorated: inner beauty is supposed to correspond to natural beauty. The Ainu strive to respect the harmony of nature.

Within this animistic conception—*anima* meaning soul, in Latin—the Ainu see a hierarchy of spirits, dominated by the spirit of fire. The spirits of the salmon and of the bear are also of great importance.[6]

The bear, lord of the forest, was the object of a special ritual after the hunt. Its spirit was to be placated by offerings of food and drink. The bear could also be killed as a sacrificial victim, after having been raised and breast-fed by a woman. The bear had to give its life so that its soul would be free of its body and could fulfill its role as a messenger to the God of the Mountain, the protector of the Ainu.

Some North American tribes, the Assiniboines, for example, perform rituals devoted to the bear; and even in Europe, in Spain (Catalonia) and France (in the Pyrenees) one finds similar ceremonies. In spite of their geographical distance, there is a similarity of modes of thinking. It would appear that Reinhold Messner was strongly attracted by the modes of thought and living of those people living north of the Himalayas.

Ainu bear festival, 1914.   PHOTO: Author's file

---

1   *La Lutte pour les Troglodytes* [The Struggle to find the Troglodytes] is the first part of a book written by B. Porchnev jointly with Bernard Heuvelmans, author of the second part: *L'Enigme de l'Homme Congelé* [The Mystery of the Frozen Man]. Both parts appeared under the title *L'Homme de Néanderthal est toujours vivant* [Neanderthal Man is still alive!], 1974.
2   B. Porchnev, op. cit., p. 146. See also O. Tchernine, p. 137.
3   B. Porchnev, op. cit., p. 146. See also O. Tchernine, p. 138.
4   Cf. Mario Mercier, *Chamanisme et Chamans.*
5   Piers Vitebsky, *Les Chamanes,* p. 61.
6   These beliefs were discouraged by both the Japanese and the Soviets.

16. Kazakhstan

In 1997 we find Messner in Kazakhstan, a country famous for its horsemen—no connection with Russian Cossaks of course. The Kazakhs are nomads living in yurts and ranging over a vast territory between the Caspian Sea in the west and China in the east.

As with many other traditional people, the Kazakhs believed in spirits. Some of these beliefs still surface today. For example, a guest may be invited to bless the lamb cooked in his honor because it is necessary to obtain from the spirit of the animal permission to eat its flesh.

Most Kazakhs are Muslims, although there are many orthodox Christians in the country. The term *albast*, close to *almas* and *almasty*, is used by Kazakh nomads to refer to the wild man. *Shaitan*—satan, devil, demon—is also used to describe the yeti by Muslim people throughout the former USSR. The Kazakhs and their neighbors the Kirghiz also use the term *ksy-gyik,* meaning "wild man."

In 1907, V.A. Khakhlov, a young Russian zoology student, heard about the ksy-gyik from his Kazakh guide. At the time he was exploring the area of Lake Zaisan, near Sinkiang (Xinjiang) and Chinese Mongolia. The young man interrupted his studies at the university for two full years to explore the wild country around Lake Zaisan and the Tarbagatai Mountains (3000 meters [10,000 feet]).

Khakhlov sought official approval and funding to mount an expedition into Sinkiang. He hoped to return with the head and limbs of a ksy-gyik. Lack of official support, followed by the beginning of the war in 1914, rendered his efforts futile. Nevertheless, a great many reports from a variety of travelers suggested the presence of a strange creature in Central Asia.

One of these travelers was Prjevalsky, a colonel in the Russian army, whose name has become famous as the discoverer of a race of wild horses, covered with thick hair and, with their large heads, looking like wild asses. This primitive relative of modern-day horses was baptized *Equus prjevalskii*. It was during his first expedition that Prjevalsky heard about the *khoun-gouressou,* meaning the "man-beast." He offered a bounty to whoever would bring him a specimen. Someone brought him a stuffed bear and the explorer concluded that the khoun-gouressou was a variety of local bear.

During his third expedition, in 1879, one of his riders was chasing a wounded yak when he found himself face-to-face with a group of hairy wild men. Prjevalsky made no mention of this in his official report, but Khakhlov heard about it from one of the members of the colonel's expedition. He also gathered further significant reports from nomadic Khazakhs.

One of these reports speaks of a female captured by peasants. Her body was hairy, the chest narrow, the head deeply set in the shoulders, the arms long and dangly, the legs short. Her feet were large, with widely spread toes, but her hands were long and narrow. Her posture was stooped:

> She was still young, completely hairy and speech-less...whenever a human being approached her, she would yelp and show her teeth. During the day, she would sleep in a position often adopted by small children: the "pose of the child" or "like a camel" in the words of an eye-witness, leaning on her elbows and knees, with her forehead on the ground and her hands on the nape of her neck. Thus, she had calluses, like the "sole of a camel" on her knees, elbows and forehead.[1]

Khakhlov noted that the Kirghiz called the wild camel *t'ë-gyik*, wild horses *at-gyik,* and wild men *ksy-gyik*. He concluded that these different creatures must occur in the same geographical area. His strong interest in local names contrasted with the attitude of some "official" scientists. Khakhlov's rich collection of information warranted a detailed report and he wrote an extensive memoir, dated June 1, 1914, to the Russian Academy of Science in Petrograd. Half a century later, Boris Porchnev went looking for the document and discovered it filed under: "Notes of no scientific importance."

Obviously, few people shared the young zoologist's enthusiasm for his "antediluvian man," in spite of Khakhlov's conclusion that: "The content of these stories told in the field by eye-witnesses is enough to conclude that they are not merely mythological or imaginary. There can be no doubt as to the existence of such a *Primihomo asiaticus,* as one might name it."[2]

Another report originated from a Russian geologist, B.M. Zdorik, working in the Sanglakh Range, part of the Hindu Kush mountains. Having asked for a list of the local fauna, he was told of wild boars, wolves, bears, hyenas, porcupines and jackals. Later however he heard the word *dev* or *deva*, a word meaning "impure spirit," which he had heard in other regions of Tadjikistan. The term is of Indo-European origin and is the root of the English word "devil."

When quizzed, the local chieftain said that dev were occasionally encountered, singly or in pairs. The Tadjiks had caught one that stole flour and grain from a mill. After a couple of months, the creature had managed to break its bonds and escape.

Zdorik lived for many years in Tadjikstan. In 1934, he was trekking on a high plateau (2800 meters [10,000 feet]) with a local guide when he suddenly came upon a dug-up part of the path, as if worked with a shovel. On the ground, there were blood spots and tufts of marmot fur. A creature was stretched out on the ground; its body was covered with coarse hair closer to that of a yak than to a bear's. The guide immediately pulled on Zdorik's sleeve and told him to run away as fast as he could. Describing their fear, Zdorik wrote:

> Never before, had I seen such an expression on a man's face. His fear communicated itself to me, and beside ourselves, without glancing backwards at the creature, we both fled away down the path, enmeshing ourselves and stumbling about in the high grass.[3]

The villagers told them that they had just stumbled upon a sleeping dev, an animal that dwelled in the mountains and with which they were quite familiar. Such an encounter was however an ill omen.

Without hesitation, Prof. Porchnev followed in Zdorik's footsteps. He had already gathered testimonies from equally reliable specialists (geologists, engineers…). In 1961, he set on a reconnaissance trip to Central Asia, where he found that the information arising from Tadjikistan was most promising. Having heard about

Professor Porchnev.
PHOTO: Author's file

the sleeping dev, thought to be an animal by the people of the region, Porchnev wondered whether it was a bear or an apelike creature. He found from a pair of engineers particularly interested in the wild man a tidbit of information about Asian pharmacopoiea. There exists a medicine fabricated from the grease of "wild man," a generally rare and expensive product, except in a certain village located in a zone where these hairy beings were reputed to be abundant.

One immediately thinks of bears, whose body parts are traditionally used in Asia as balms and ointments for various reasons, notably as aphrodisiacs.[4] Even today, bears' penises are found in jars in traditional Chinese medicine pharmacies. Porchnev noted the name of the remedy, *moumieu*, from the Iranian *moum*, meaning wax or grease; he also noted its resemblance to the Tibetan *mi-gheu* for wild man. In an area suggested by Zdorik among others, Porchnev bought some *moumieu*. It was unfortunately of the mineral variety, useful but concocted from petroleum products and unrelated to wild man grease. However, that area was home to a rich fauna: bears, wild boars, lynx, wolves, foxes, martens, badgers, porcupines and marmots, and a luxuriant flora, in some places impenetrable. Stream banks and summits remained inaccessible to Porchnev's expedition.

The abundance of berries, fish and game create an ideal refuge for a population of large creatures, especially in combination with its remoteness from human settlement—a guarantee of the presence of large animals…or wild men. Nevertheless, Porchnev felt that obstacles were being put in his way. Time and again, the evidence evaporated, as it did for the moumieu He blamed his difficulties on the role played by local religions. A map of the distribution of ethnic groups shows that it overlaps with those regions where Islam, Buddhism and Shamanism are widespread; there are, in addition, a few local "pagan" cults. Porchnev came to the conclusion that over the past few millennia, perhaps the last few centuries, Neanderthals had survived in those regions where they were protected by religions or superstitions. For example, leading lamas have issued a special edict protecting the last mi-gheu. Islam, originally spreading at the expense of Zoroastrianism, has become, by default, the protector of the dev:

> Specific interdictions and instructions have been issued to the true believers about the near-men, those "spirits" strangely material as well as mortal [5]

Religions play an important role in the countries visited by Reinhold Messner. They constitute a factor that just cannot be ignored. In some cases, wild men have certainly been protected; in others they have been hunted, again for religious reasons. It becomes very important to understand the cultural history of the regions visited, which requires patience and subtlety. The dividing line between physical reality and metaphorical speech is often vague, making it difficult to know when natives speak of the world of spirits and that of *near-men, strangely material as well as mortal.*

Depending on the attitude of the investigator, the religious dimension will appear either as an obstacle, a veil over the phenomenon under study, or as a component which must be included to understand it. In that case, religious influences open the door to a beginning of understanding and help lift the veil obscuring physical reality.

1　Boris Porchnev, op. cit. p. 53.
2　V.A. Khakhlov, as quoted by Boris Porchnev, op. cit., p. 59.
3　Odette Tchernine, *In pursuit of the abominable snowman,* p. 48.
4　I have seen such jars on the shelf in an "old-style" pharmacy in Chengdu, Szechuan in 2008. I deplore the hunting of bears, whose numbers are in alarming decline.
5　Boris Porchnev, op. cit., p. 161.

17. Mongolia

In 1998, Messner traveled to Mongolia, a country where the faithful of many religions have had to submit to the communist regime. Today, 50 percent of Mongolians are Tibetan Buddhists, 6 percent shamanists or Christians, 4 percent Muslims and 40 percent agnostic or non-practicing.

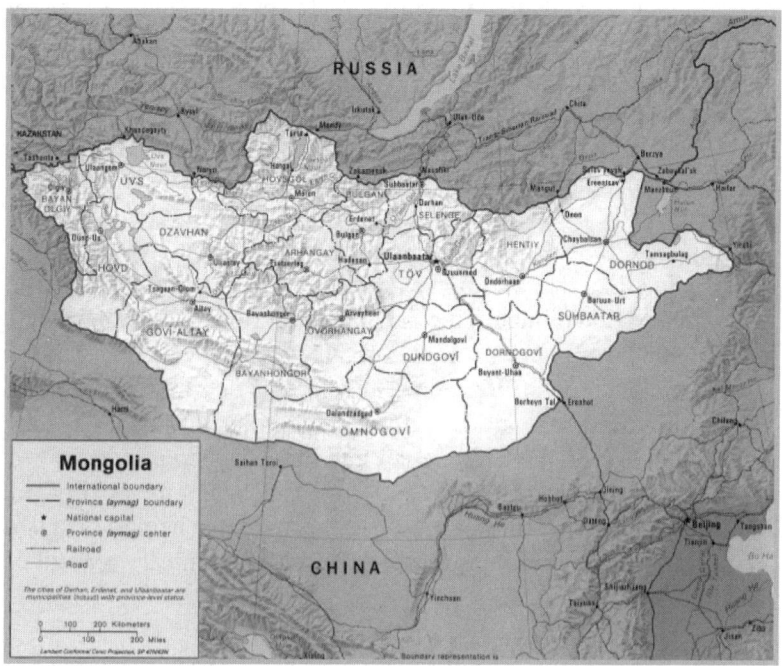

Mongolia. Map: Wikipedia

Already in the seventeenth century the lamas attacked the shamans; the "yellow faith" of the "yellow hats" of Tibetan Buddhism faced the "black faith" of the shamans. Shamanisn was, however, so deeply entrenched that what followed was a mixture of both religions. This syncretism is still reflected in Mongolian Buddhism today.

Once again, ignoring the shaman means going without a potential ally, especially among traditionally nomad or semi-nomad populations: "In him are united religion, psychology, medicine and theology, all of which are separate areas in the West."[1]

Do shamans possess a special ability to get along with wild men? Can they approach them easily? Even dominate them? Do they take advantage of them or of their remains for ritual purposes?

Neither shamans nor lamas seem to fear encounters with the wild man, specifically the Mongolian *almass*. Professor Baradyine, orientalist and explorer, relates how in 1906 his caravan set up camp for the night. At the top of a sandy ridge there appeared the simian silhouette of a hairy man. The professor asked his guides to go after the creature, but only a lama ran towards the almass, declaring that he felt sufficiently strong to immobilize it. However, the almass quickly disappeared behind another dune. The Mongolian members of Baradyine's party pointed out that it was as rare to encounter an almass as it was to run into a wild horse[2] or a wild yak.

The Mongols call the wild horse *tahi*. Until recently, official zoology considered that such local names referred to legendary creatures, but there really exists a tahi. Odette Tchernine provides important information: "It became evident to the nomads of this region that the Almass, just like the wild horse and the wild camel, were avoiding the neighborhood of man. They had moved further and further away, as the people were extending their grazing areas."[3]

On the other hand, Boris Porchnev points out the wide-ranging distribution of almass in Mongolia, although already by 1927 they were found only in the Gobi Desert and in the Kobdo province.

Another prominent Orientalist, Prof. Rinchen, a member of the Mongolian Academy of Sciences, gathered with the help of his team a large number of eyewitness reports. Here's his summary portrait of the almass:

> The Almass are very similar to people, but their body is covered with reddish black hair which is not very dense: their skin remains

visible through the hair, which never happens among the wild animals of the steppes. Their stature is similar to that of the Mongols, but they are slightly stooped and walk with their knees half bent. They have massive jaws and a low forehead. Their brow ridges are quite prominent compared to those of the Mongols."[4]

The Mongols speak of an ill-kempt individual as "bristling like an almass."

In 1937, during a Japanese offensive in Outer Mongolia, Soviet soldiers shot down two individuals who hadn't responded to a call to lay down arms. They found that the victims were very apelike. Their interpreter, an elderly Mongol, told them that these wild men were sometimes seen in the hills. As tall as a man, they were covered with moderately dense reddish hair. Their thick hair hid their forehead but not their bushy eyebrows.

Twenty years later, a considerable number of eyewitness reports about the almass had accumulated. In the Gobi Desert in particular, juvenile almass had been seen, alone as well as with their mothers. Porchnev reports on the ethnographic enquiries of Prof. Rinchen and his colleagues, expressing his strong admiration for his work:

Prof. Rinchen (right) with Polish colleague, W. Plawinski, 1967.
PHOTO: Author's file

—a magnificent elder, sporting enormous Mongol-style drooping mustaches, always garbed in the colourful national frock. Extraordinarily erudite, he has assimilated various western cultures as well as the Russian and the Mongol, and is considered a world-class orientalist…As early as 1958, Rinchen published in *Sovremennaya Mongolia* (*Contemporary Mongolia*) an article entitled: "A Mongol ancestor of the Wild Snow Man?"[5]

Once more, we must acknowledge the quality of the witnesses: hill people, shepherds, soldiers and officers, explorers, scientists. The precision of the evidence stimulated further work by researchers and everyone eager to gather more information on the creature that a witness, a marshal of the Soviet Army, described as "a fossil ape-man."

A report that stands out among the many collected by Rinchen is that of the body of an almass nailed to the ceiling of the Baruun-Urt monastery, in northeast Mongolia. According to the witness, reporting around 1937, the spread-out limbs of the relic were similar to human arms and legs. The dark face was partly obscured by long dangling hair. The skin of the body was decorated by paintings and scattered mystical incantations, the work of lamas. The witness could not recall the presence of hair on the body. This almass, exhibited in a monastery, is reminiscent of Reinhold Messner's discovery of the skin of a "red yeti" in a Bhutan monastery.

1 Piers Vitebsky, op. cit., p. 154.
2 Prjevalsky's horse.
3 Odette Tchernine, op. cit., p. 53.
4 Boris Porchnev, op. cit., p. 43. B. Porchnev judges these reports reliable: "These carefully worded documents were the results of a long and dedicated research by the emerging Mongolian scientific school, before it split up in a variety of individual disciplines."
5 B. Porchnev, op. cit., p. 45.

18. Diogenes in the Himalayas

On the roof of Asia, life is constantly in close contact with spirits. The natives' existence is replete with symbols, rituals, prayers, meditations and dialogues with the world of the living. While religious activity is most intense in the monasteries, one should not forget the numerous sedentary or wandering hermits, Himalayan versions of Diogenes, living in caves, in the forest or in abandoned houses.

Alexandra David-Néel, the veteran traveler who long lived in Tibet, spent a whole winter as an ascetic, living in a hut of rough-hewn timbers backing onto a cave. She ate but a single meal a day.

> Both the spirit and the senses are sharpened through this life style, constantly filled with contemplation, observation and reflections. Does one then become a visionary; or rather has not one remained till then completely blind?[1]

Given the mystical context of the country, one should not be surprised to hear that she wore, as part of her hermit's garb, a rosary made up of 108 discs cut out of 108 human skulls, a magic dagger at her belt and a trumpet carved out of a human femur. That a lama called her the "Reverend Lady" is a sign of the respect given to a woman who held a high rank in the Lamaist orders. The lama lady had been fully initiated; she even occasionally resorted to witchcraft. She was respected and also somewhat feared.

The prehistoric appearance of the hermits' refuges is not surprising. They are mere caves, closed by a wall made of stones, dirt and turf, and entered through a curtain of yak hair. Some hermits even survive naked in snow-covered hills. How can that be possible?

It is well known that Tibetans can endure very long hikes: walking steadily for 24 hours is by no means a record. The loung-gom-pa lamas are initiated through a series of breathing exercises before learning a mystical formula; they focus their thoughts on a rhythmical mantra that guides their breathing during a trek.

> Under the trance state, although much of normal consciousness is abolished, there remains enough to lead the walker towards his goal and to avoid obstacles he might meet along the way. However, this is done entirely mechanically, without the need of conscious thought on the part of the walker in a trance.[2]

What is even more surprising is survival in the mountains at 4500–5000 meters (13,000–17,000 feet) without succumbing to the cold. Such a feat also requires teaching by a master: the key is to learn to stimulate the internal warmth called toumo (see also, chapter 7). Alexandra David-Néel went through this training and, at the request of a lama, successfully bathed in the icy waters of a river and, without getting dressed, meditated for a whole night: "It was at the beginning of the winter, at an altitude of about 3000 meters (10,000 feet). I felt enormously proud of not having caught a cold."[3]

Learning about toumo is a practice related to the Hindu hatha yoga. The apprenticeship is lengthy and complicated, requiring patience, concentration and application. Both practices, sustained walking and the marvelous production of internal warmth, show surprising aspects. For example, the trek of the loung-gom-pa walker must not be interrupted by speaking to him or by taking his picture. The walker is in a trance and a nervous shock might be dangerous, even lethal.

As to the toumo master, he runs no risk as long as he respects the rules—which are not designed for people with weak lungs! Competitions are held to measure the strength of toumo champions. For example, a competitor sits in the snow: the radius that the snow melts around him is a measure of the amount of heat he generates.

Isn't there some link between the rhythmic pace of the loung-gom-pa lamas and the pace of the wild man? Is it outrageous to suggest some similarity between the two modes of motion? The mechanical pace, the speed of displacement, the ability to follow a path without detour, are also characteristic of the movement of the unknown creatures reported in Asia.

Further, the size of the footprints, sometimes out of proportion with the bulk of the creature, suggests a strong heat flow: the snow has melted, expanding the size of the footprint.

The suggestion that the yeti might be confused with a hermit was put forward as early as 1935. A few years later, in 1958, the German missionary and physician Father Franz Eichinger stated that the abominable snow men were actually hermits living in the upper reaches of the Himalayas. They lived as the Christian saints used to and devoted their existence to prayer and the healing of the sick.

The assimilation of hermit and yeti is also mentioned in some tales gathered by Nepalese historian and folklorist Kesar Lall. In one of these stories, a lama tames a yeti and shelters it in his refuge. The yeti hunts and shares the meat he catches. When the yeti dies, victim of an avalanche, the lama cuts up its body to feed the vultures, as tradition demands. The lama then preserves the head of the yeti, supposedly still shown today at the Pangboche monastery.[4]

In 1980, a group of young women skiing among the high peaks in neighboring India encountered a holy man, naked, unaffected by the cold. An Indian journalist then commented that the word yeti was derived from *yati*, a hermit, a being free from all social bonds, without distractions and material protection, as pointed out by Kesar Lall.

There is thus clearly a relationship between religious man and wild man.

The wild man is widely feared and is to be avoided. Just seeing it brings bad luck. Sometimes the witnesses are subject to uncontrollable bouts of terror and die within weeks of the encounter. One should remember that the yeti is also called *dremo*, a demon in Tibetan. For Alexandra David-Néel, Tibet is indeed the land of demons. Numerous and complex rites govern relations with these demons.

Today, in spite of political turmoil, Tibetans and a few visitors continue to frequent the sacred places. The Yerpa

North European (sami) shaman, 1767.
Illustration: Public domain

Valley draws pilgrims, philosophers and poets, and all who are attracted by the sacred places of meditation. Visitors acclimatized to the altitude (4000–4500 meters [13,000–15,000 feet]) recognize in the men or women walking clockwise around a *chorten*, cranking a prayer mill, fellow seekers of inner peace; language barriers vanish.

The Yerpa Valley, about 16 kilometers (10 miles) north of Lhasa, is peppered with caves that used to shelter hermits. Some date from the days before introduction of Buddhism. Caves are hallways to spirituality and continue to be seen as important centers of meditation and spiritual energy. Tibet is still the land "where the inhabitants believe in demons and Bodhisattvas and where Dakinis walk across the sky."[5]

This transition from the world of the living to that of the gods, from the material to the spiritual, is most unsettling to our rationalist attitude, especially since that transition takes place without warning, as part of the everyday life of Tibetans. To draw a boundary between the concrete and the invisible would not make sense to them. For example, materializations, to take one phenomenon among many others, are of interest but do not trigger enquiries, interrogations or efforts at "scientific" explanations. Phenomena that to us may appear extraordinary do not upset Tibetans' "accepted ideas about the laws of nature and what they say regarding what is possible and what isn't."[6]

The old Bon religion, as well as the different forms of Buddhism, seem extraordinarily complex to western eyes. Their concepts are to be absorbed gradually, over sufficient time. The yeti's existence deserves to be considered equally from the perspective of Tibetan mystical practices.

There remain within the human soul unknown regions, as there remain unexplored areas of the world. Some lands vibrate with the voices of the gods: the mountains of Tibet, the great forests and deserts. In Tibet, there are as many gods whose presence may be felt as there are high peaks. Sometimes the gods are expelled by brute force and the sacred places are destroyed. Villagers, travelers and pilgrims then place branches festooned with colored banderoles fluttering in the wind, so as to honor the god who protects the pass or the valley. Villagers seek those "secret valleys," refuges of persecuted humans and their gods.

Of course, material evidence is particularly important in attesting the survival of the wild man. However, some data may be interpreted as part of religious elements. Until the 1930s, remains of almass were to be seen in some Mongolian monasteries. Odette Tchernine reports the words of the old man who saw the remains of an almass fixed to the ceiling of the Barun Hüre monastery. The same old man told her that in another monastery there lived a lama famous for his learning. He was called "Son of the Almass," for his father, who was also a learned lama, had been captured by almass. The son was born of his union with a female almass. Having escaped with the child, the father and his son were picked up by a caravan. The son also grew up to be a learned man.

One last look at Tibet: Reinhold Messner's travels have served as a pretext for our search for the wild man in neighboring countries. Let us now return to the land where he started his quest. As we have seen, the information that he gathered in Tibet shows a strong resemblance to what he found in neighboring countries.

In the spring of 1991, as we recall, Messner was on a yeti expedition in the company of three members of a German television station. He had stopped to view the carcass of a yeti at the Gangtey Gompa. A closer look led him to conclude that the hands as well as the legs came from the body of a child aged eight to nine years. However, a mask took the place of the face. As Messner explained:

> Over there, the peasants are descendants of Tibetan nomads whose religion—Lamaism—is filled with tantric imagery. Their beliefs, as those of shamanism, joins all aspects of nature—water, air, wind, clouds, fire and love—to the divine. Without doubt, the Yeti belongs to that animistic vision of the world.[7]

The visit to that monastery left Messner puzzled. Why would the erudite lamas preserve a myth of shamanic origin so foreign to the rational world? Could it be an antithesis to human beings, a symbol of brutality at the antipodes of civilization? Messner couldn't figure it out. He concluded that it was only in such Himalayan lands that the yeti could have survived.

In the French version of his book, Messner shows a picture of two carcasses, one being a yak, the other a chemo, both hanging from the ceiling near the entrance of the Sosar Gompa. The chemo with his fierce teeth and claws is easily capable of killing a yak and tearing it to pieces. According to Messner, that chemo, or dremo, or yeti, is a kind of particularly powerful brown bear.

That would be a plausible conclusion outside the context of the ancient religion: in this case, the yeti is simply a zoological entity. However Messner realizes that he is dealing simultaneously with an animal and a myth:

> In these lands, where to everything there corresponds another half—as per the principle of yin and yang—the yeti is also seen as a dual being, each half of which, the mythical and the real, is truly alive.[8]

Messner attributes a particular importance to the antique religious phenomenon, going perhaps all the way back to the prehistory of the Himalayan region. Shamanism is at the very core of this tradition and should not be seen as a superficial phenomenon, and we should thank Messner for drawing our attention to shamanic practices. Perhaps he knows more than he tells us? Why would he emphasize discoveries that turned many of his admirers into ferocious critics? We strongly recommend Messner's work to the reader: it is the account of a traveler who has crossed Nepal, Tibet, Bhutan, China and Mongolia, traveled among the Kirghiz and Kazaks, as well as in India and Pakistan. Many readers have been struck by the strength of his work and by the aura of truth that surrounds it. Messner's words are in harmony with those of the investigators who came before him and whose works are fundamental to understanding those countries.

---

1 Alexandra David-Néel, *Mystiques et Magiciens du Tibet*, p. 85.
2 Alexandra David-Néel, op. cit., p. 226.
3 Alexandra David-Néel, op. cit., p. 232.
4 See Kesar Lall in *Lore and Legend of the Yeti*, p. 3.
5 Alec Le Sueur, *Running a Hotel on the Roof of the World,* p. 199.
In Tibetan Buddhism a bodhisattva is anyone who is motivated by compassion and seeks enlightenment not only for him/herself but also for everyone. A Dakini is a female embodiment of enlightened energy, dancing across the sky.
6 Alexandra David-Néel, op. cit., p. 255. Her words remain as relevant today as when she wrote them.
7 Reinhold Messner, *My Quest for the Yeti,* p. 100.
8 Reinhold Messner, op. cit. 100.

This remarkable painting of a yeti by the noted Canadian artist Robert Bateman is possibly very close to what the creature actually looks like. There are no photographs of the yeti, so any images are based on descriptions by people who have claimed to have seen it. Bateman is one of the most, if not the most, famous naturalist artists in the world—his insights into the possible likeness of a yeti are significant. PAINTING: © Robert Bateman

In 1966 Bhutan issued these stamps showing five different views of the yeti. The stamps reflect both mythological aspects and aspects of reality. There can be no doubt that the creature is deeply woven into the culture of the region.
PHOTO: Author's file

This intriguing illustration for the yeti appears to be very old. The third image is said to show the way in which the creature sleeps. This rather odd sleeping posture has become common knowledge, so we might conclude that the creature has been actually observed as it slept.
PHOTO: Author's file

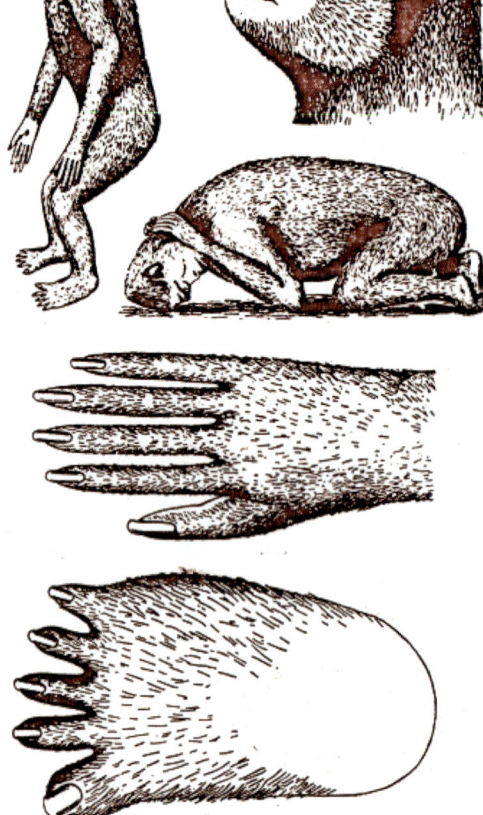

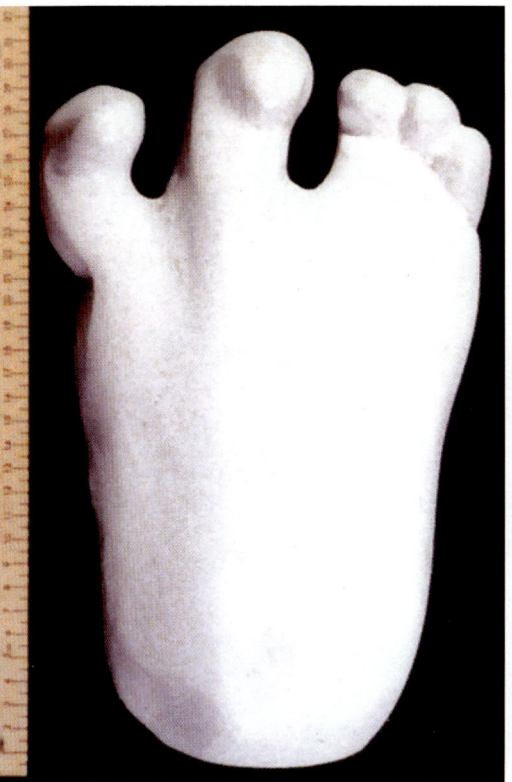

A plaster cast (copy) of what is believed to be a yeti footprint. The print was in snow and the cast (original) was made from a photograph. Given this is the nature of a yeti's foot, it can be assumed that the creature is quite different from other Asian hominoids. The cast measures about 12 inches (31 cm) long.
PHOTO: C. Murphy

One of three alleged yeti scalps. One of them, likely this one, was sent for professional analysis and determined to be made of the skin of a serow (member of the goat-antelope family). The other two scalps, said to be 350 years old, have not been examined. Nevertheless, it has been assumed that they were likely of the same nature. Possibly one of the two is authentic. Certainly, if one monastery had a scalp, then others would want one also, which could lead to a scalp (or several) being fabricated. In time, they would all take on the same significance—that of a genuine relic. Photo: Fortean Picture Library.

An alleged skeletal yeti hand. Analysis of the hand in the late 1950s and 1960s was inconclusive because agreement was not reached by the several professionals who analyzed it. Modern methods would probably provide firm results, but unfortunately the hand was stolen.
PHOTO: Fortean Picture Library

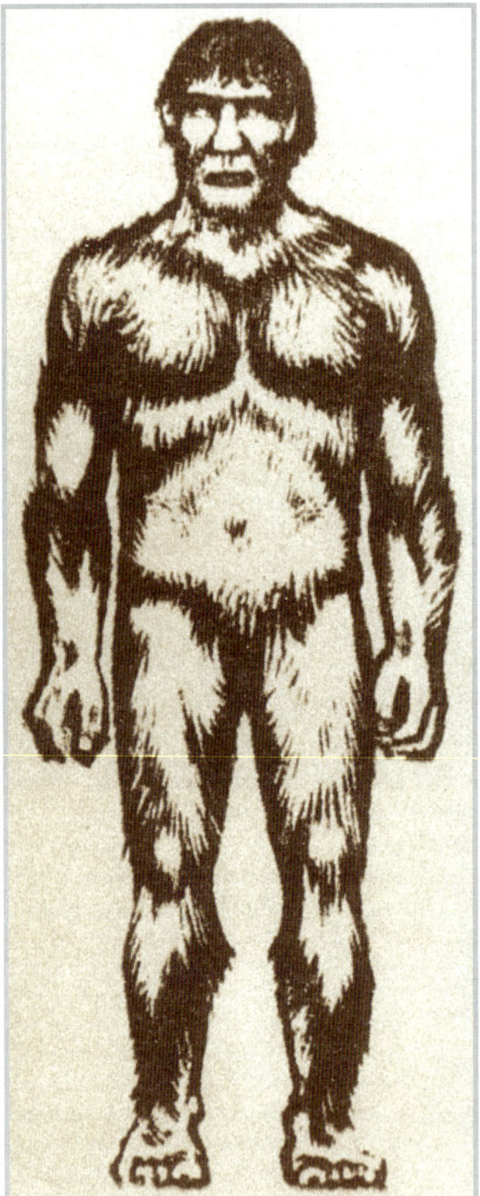

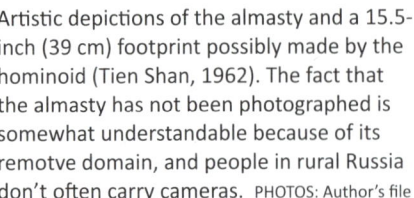

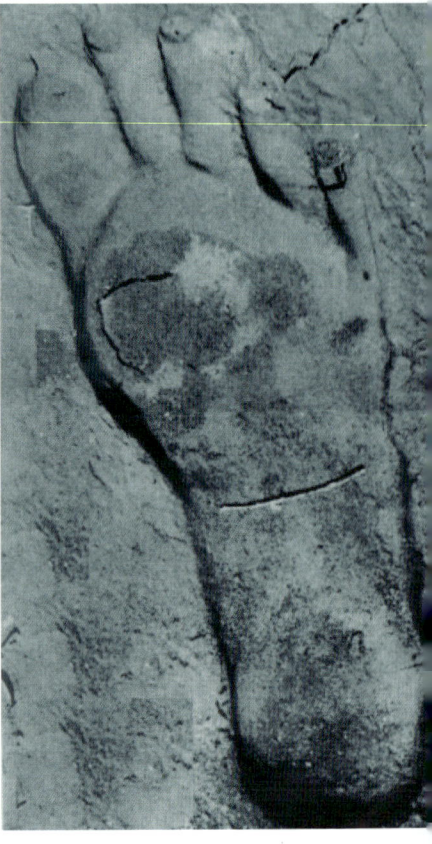

Artistic depictions of the almasty and a 15.5-inch (39 cm) footprint possibly made by the hominoid (Tien Shan, 1962). The fact that the almasty has not been photographed is somewhat understandable because of its remotve domain, and people in rural Russia don't often carry cameras. PHOTOS: Author's file

Dr. Marie Jeanne Koffmann carefully lifting a glue-treated track which is shown below with a ruler. It measured about 10 inches (25 cm) long and was found (one is a series) in the Dolina Narzanov Valley, North Caucasus in 1978.

Generally speaking, the almasty is not an exceedingly tall creature like North America's sasquatch. However, in both cases human-size tracks have been found which are likely those of younger creatures or smaller adults.

In the 1800s and first part of the 1900s, the almasty was often seen in rural Russia. Some people had great compassion for it and left out vegetables and other food items for it to take. These people were not aware or concerned with the scientific implications of the creatures' existence so did not obtain any physical evidence. Although there are still sightings, it is believed the almasty population has greatly diminished.

PHOTOS: I. Bourtsev

Igor Bourtsev, a prominent Russian hominologist, is seen here comparing a casts of a possible almasty footprint to his own foot. The cast measured about 14 inches (36 cm) long. The print from which the cast was made was one in a series found in the Pamir–Alai Mountains (Tajikistan Republic) in 1979. Boutsev was heading an expedition in this region and the group discovered the footprints one morning about 70 feet (21.4 m) from their tents. PHOTO. I. Bourtsev

Zana, the Russian ape-woman (or almasty) is seen here in this remarkable artwork by Branden Bannon. Zana was captured, trained to do simple chores, and later abused by several men during wild drinking bouts. She had "normal" babies, four of whom survived to adulthood. Zana died in the 1880s or 1890s. People where she lived remembered her when they were questioned in 1962.
PHOTO: B. Bannon

Igor Bourtsev examining the skull of Khwit (seen on the right), Zana's youngest son (1964). Bourtsev attempted to find Zana's remains for scientific evaluation, but was unable to locate her grave. He thereupon exhumed the remains of Khwit (died 1954; grave was definitely identified) for the same purpose.

Russian scientists determined that Khwit's skull was different from that of ordinary humans, however, the American scientist, Dr. Grover Krantz, said that it was that of a fairly normal human being.

DNA (obtained from a tooth) indicated that Khwit was human. This whole issue begs the question: Could the almasty simply be a different type of human? PHOTOS: I. Bourtsev

An old Chinese drawing of what is believed to be yeren. The writing in the top right hand corner states: "Xin Xin is small and likes to bark. [Xin Xin] lives in the mountainous ravines, resembles an ape, has human face and limbs, head hair is long, the head and face are put straight. Its voice is like the crying of an infant and the barking of a dog."
PHOTO: Public Domain

A contemporary depiction of the yeren. There have been some unusual stories about this creature. One such story reported in the *World Journal* (Taiwan paper, October 1997) tells of a woman who stated she was abducted by a "wild man" (assumed to be a yeren) and had its child. Video of the child, at age 33, indicated that he had a small head and what appeared to be a kind of tail. The article went on to state that his body shape, arms, and legs, were similar to those of the North American bigfoot. However, he did not have any noticeable long hair. It was stated that the Chinese wild man has been recorded as far back as 100–200 BC, also that a "monkey-boy" was discovered in 1932, but its existence was not reported until after it had died. PHOTO: Author's file

Dr. Zhou Guoxing, the noted yeren authority, is seen here (center) with Dr. Grover Krantz and his wife, Dian. Dr. Krantz, who was highly involved in sasquatch research, went to China in 1995 to explore the yeren question. Like the yeti, almasty, and the sasquatch, firm scientific evidence of the yeren's existence has never been obtained. PHOTO: G. Krantz

19. Looking Back: Marie-Jeanne Koffmann

We have drawn repeatedly on the writings of poet and essayist Odette Tchernine in the 1970s. It is easy to forget that she was one of the first to write seriously about the wild man in Russia and Asia. She introduced the works of Professor Porchnev and Dr. Marie-Jeanne Koffmann to English-speaking readers. Her spirited style and the wealth of information she presents are widely acknowledged. English poet John Heath-Stubbs paid homage to her in a poem entitled *The Yeti:*

> "Brother," he grunted, "who have called yourself
> Sapient, and me abominable —
> Your sapience is the knowledge of good and evil.
> My breakfast and my lunch are mountain lichen,
> Or sometimes I can catch a calling-hare;
> But never took a bite of that apple.
>
> Well, when you have torn yourselves apart,
> And split the world in two, we will be standing,
> Ready to take over — and at the door of history, there waits
> Another Eden, the same poison-tree."[1]

This poem brings to mind Marie-Jeanne Koffmann, an example of Umberto Eco's aphorism "The history of scientists offers more romantic possibilities than that of ordinary people."[2]

The life of Marie-Jeanne Koffmann is a real novel. Born in Paris in a French family, she joined her father who was actively involved in the Russian revolution. At the age of 18, she became, unwillingly, a Soviet citizen. During the war, she led a battalion of mountain troops in the Caucasus, and then became a researcher at the Academy of Medicine of the USSR. Her double citizenship was enough to attract the attention of the KGB and send her to the Gulag for seven years. Regaining her freedom, she became a general surgeon in Moscow and participated in expeditions in the Pamir Mountains where she developed an interest in the snowman. As a member of Russian learned societies, Marie-Jeanne Koffmann benefits from a pension: 120 rubles a month, the equivalent of three French francs (1992; less than one Euro!). In recognition of her double citizenship, France has awarded her the RMI,[3] which she found "splendid."[4]

This short biographical notice was part of an article about the 1992 Franco–Russian expedition of which Marie-Jeanne Koffmann was the scientific leader. A team of videographers, under the direction of producer Sylvain Pallix, accompanied the expedition and produced a documentary: *Almasty. Yeti du Caucase* (TV Channel France 3, February 13, 1993).

The expression "Yeti of the Caucasus" seemed appropriate in view of the close relationship between the two snowmen, and brings to mind Odette Tchernine's comment: "Local folk knew in former days all about the Almas, or Yeti-Snowman; the majority of scholars and officials in the towns did not."[5]

One should however look closely at what is meant by "snowman" in the Russian–Asiatic context. Marie-Jeanne Koffmann has repeatedly asserted that the yeti is an anthropoid ape, a relative of the orangutan living in the high forest of the Himalaya, and NOT in the snows. It would frequent snowy areas only while passing from one forest to another.

In spite of her extensive investigations in the 1960s, Marie-Jeanne Koffmann never managed to see the

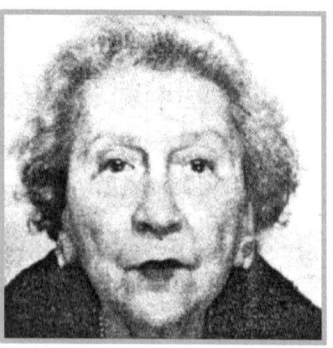

Odette Tchernine, member of the Royal Geographical Society.
PHOTO: Author's file

snowman herself.[6] Odette Tchernine summarizes Koffmann's research:

> "Doctor Koffmann's field work in the Caucasus suggests by her down-to-earth eye-witnesses' reports, and the careful evidence of her own team, that the Almas, Snowman, Yeti, "wild man" is there, and is neither legend, hallucination, nor Shaitan."[7]

Witnesses of all ages reported sightings. One of these came from an old man who was keeping an eye on a field of sunflowers that had been repeatedly visited by an intruder. This field bordered on a river, near Nal'chik, the capital of Kabardino–Balkaria. The old watchman fired his gun and scared off a stocky almass of medium stature, which ran away, crushing sunflowers in his flight.

In their research, Marie-Jeanne Koffmann's team found two dens and a "pantry" holding two pumpkins, eight potatoes, a half-eaten corn cob, the remains of a sunflower, blackberries and three apple cores. Not to forget four horse droppings, which suggest that the almasty enjoy these excrements because of their salt content. This short inventory suggests that the almasty of the Caucasus are mostly vegetarian.

Koffmann also remarked on the curiosity shown by the almasty, who do not hesitate approaching new objects—tents, clothing, utensils—and to manipulate them under cover of darkness.

Kabardino–Balkaria, a constituent republic of the Russian Federation and the home of Kabards and Balkars, is a rather special region where almasty live in the proximity of villagers. The country is a sort of field laboratory for wild-man research. One particularly interesting report is that of a Russian zoological technician who worked in the area in 1956. She roomed in the house of an employee of the kolkhoz. One evening, during a wedding celebration at the neighbor's house, she was resting in her room with the door open. Suddenly, hearing a sharp cry, she noticed a hairy creature, squatting in the middle of her room, staring at her. She yelled, "Where did you come from?"

The creature rushed out, leaving behind a fetid smell. The next day, the Russian woman was told that it was an almasty that lived in the house next door, formerly owned by an old woman who had tamed it. Odette Tchernine thinks this anecdote has "a curious dream-like quality about it, a connotation with fables of monsters, phantoms and the like…"[8]

Marie-Jeanne Koffmann gathered hundreds of more classical first-hand reports. The following example has been selected because of the time of the event, the nature of the witness, and the presence of his colleagues in the police force.

The event occurred in October 1944, while a detachment of mounted police was crossing a hemp field. The witness was right behind the leading horseman when the latter shouted, "Look! An almasty!"

The creature stopped chewing on hemp stems and fled quickly towards a shepherd's hut. The policemen encircled the small building. They were moving in, shoulder to shoulder when the almasty rushed out, broke the circle and fled into a ravine, disappearing in the bushes bordering the river. The hirsute and robust individual, his face covered with hair, was wearing a ragged caftan. Sometimes almasty visit people's houses and are given food or some old garment.

Another interesting testimony is that of a physician, Dr. Vazghen Karapetian, cited in Dimitri Bayanov's book *In the Footsteps of the Russian Snowman*. While a military physician in the Caucasus, Karapetian examined a strange creature. The authorities suspected the individual to be a saboteur in the pay of the German army. However, he was wearing neither overcoat nor furs; his hair was his own and covered his whole body. About 1.80 meters tall (6 ft), he made sounds unrelated to any known articulated language.

Founders of Russian Hominology in 1968.

Left to right:
B.F. Porchnev,
P.P. Smoline,
A.A. Machkovtsev,
D. Bayanov and
M.J. Koffmann.
PHOTO: Author's file

The Caucasus area. MAP: Author

Following the 1960 creation of the Seminar for Research on Relic Hominoids, in Moscow, Bayanov had the opportunity of hearing Karapetian speak at a number of conferences. Karapetian emphasized three traits that distinguished the creature from human beings:

1. It easily withstood the cold, which he preferred to the temperature of a heated room. Under normal room temperature (18–20°C [64–68°F]) it sweated profusely.
2. Its face and look were expressionless, more like an animal than a human.
3. The creature had unusually large parasites, bigger and different from those found on man.

Following a detailed examination, Dr. Karapetian concluded that the individual was not a man in disguise, but a "very wild" creature and that its hair was real. Later he learned after enquiring with the authorities that the creature had been shot as a saboteur. Apparently, it had been argued in higher places that after so many years under a Soviet government, the population was civilized and that wild men had long ago disappeared. The strange creature could only be an enemy.

It also seems that after the Russian Revolution of 1917, wild men would have been collateral victims of battles between Reds and Whites. The Second World War made more victims among these vanishing creatures. It is as if the home of hairy men shrank with the expansion of civilization.

---

1   John Heath-Stubbs, *Collected Poems 1943–1987*, p. 162.
2   Umberto Eco, as spoken on France Inter, February 15, 1996.
3   The RMI (Revenu Minimum d'Insertion) was, between 1988 and 2009, a French minimum-wage allocation.
4   Pierre Berruer, *Ouest-France,* 28–29 mars 1992: "La Science traque le Yéti du Caucase."
5   Odette Tchernine, op. cit., p. 37.
6   She missed seeing one in 1959 in Daghestan when fellow investigators surprised a "kaptar" (another name for the almaas) bathing, but scared it away with a gunshot before taking a photo (cf. Porchnev, 1957, p. 168).
7   Odette Tchernine, op. cit., p. 37.
8   Odette Tchernine, op. cit., p. 161.

20. Mystery of the Braided Manes

Some recent reports are particularly interesting. For example, the following, dating from August 1991, by Gregory Panchenko, a young biologist and close collaborator of Marie-Jeanne Koffmann, who often relates his adventure during her presentations.

A curious trait of wild men's behavior is their interest in horses, and in particular their manes and tails. They are said to busy themselves at night with braiding horses' hair. Various investigators have mentioned this strange habit, including Professor Porchnev and, well before him, William Shakespeare:

> This is that very Mab
> That plaits the manes of horses in the night
> And bakes the elflocks in foul sluttish hairs,
> Which once untangled much misfortune bodes.[1]

The appearance of Queen Mab, the Dream Fairy, is a most surprising part of the play, *Romeo and Juliet*. It suggests that the author, an erudite Renaissance man, was aware of the main features of some very ancient phenomena.

Romeo and his friends are getting ready for the masked ball, part of the festivities of the Capulet family. As we all know, these maskbearers are in for a tragic fate, which brings to mind the equally tragic and lethal conclusion of another famous event, the Bal des Ardents. On January 28, 1393, the king of France, Charles VI, and members of the court frolicked as wild men, disguised in costumes of oakum glued with pitch, to simulate a hairy pelt. A torch ignited the costume of one of the courtiers and the fire spread among them all. The king was saved (by a woman who covered him with her cloak to choke the flames) but four of his guests were roasted alive.

The presence of wild men is often taken as an omen of great misfortune, then as now, in Europe as in Asia and elsewhere. The costumes of the Bal des Ardents suggest a probable survival of wild men at the end of the Middle Ages. An important chronicler of the medieval era, Jean Froissart, speaks of such hirsute beings during the reign of French king Jean le Bon (1319–1364). During the presentation of gifts to the queen, a procession of forty prominent Paris burghers paraded through the city. The presents were shown to the spectators from a litter carried "by two strong men, most appropriately dressed as wild men…"[2]

We now return to the deeds of wild men in the twentieth century, and more specifically to the report of biologist Gregory Pachenko, who couldn't quite swallow the explanations, however plausible, put forward for the braiding of manes. One zoologist had suggested, for example, that a weasel, while hunting for mice, had jumped onto the horse's back and messed up its mane. It could also happen that a horse, while shaking its head, might braid its mane, especially if the stem of some weed or a blade of grass was caught in its hair.

Panchenko was about to leave Kabardia, where he had been collecting information about the almasty, when he heard about the braiding of horses' manes in a collective farm. At the farm, he asked permission to stand on guard in the barn during the night. Above the main double door there was an open window, three meters (10 feet) above the ground. On the left, going in, a mare was tied to a feeding trough; on the right, a bed behind a blanket reached to the ground. From the bed, hidden behind the blanket, Pachenko kept an eye on the mare and on the window. Tired, he fell asleep; suddenly, the mare snorted and woke him up. By the moonlight he saw a silhouette, about 1.70 meters (5 feet 6 inches) tall, its head low on the shoulders, busy braiding the mare's mane. He watched it for five to seven minutes; the mare did not react. Suddenly Pachenko felt as if the creature had detected his presence; it jumped, as agile as an ape, onto the edge of the wall and up to the open window before disappearing into the night. Subsequently, Pachenko noticed that the braiding was rather clumsy, perhaps the work of a young and inexperienced almasty.

Commenting on Pachenko's "planned" encounter with the hominoid, another prominent Russian investigator, Dimitri Bayanov, thought that both braiding techniques were real, one being natural (the weasel, the shaking of the horse's head), the other artificial, as explained by folklore, the local traditional knowledge which attributes to the almasty a fascination for horses and the habit of braiding their manes.

Bayanov has devoted much of his life to proving the existence of wild men. He collaborated with Boris Porchnev, Marie-Jeanne Koffmann, Gregory Panchenko, Igor Bourtsev, Maya Bykova and many other "hominologists" worldwide. In his words, "The lessons of this event are momentous and have to be thoroughly learned."[3]

Bayanov's *Russian Snowman*.
PHOTO: Dmitri Bayanov

Shakespeare brings to the stage a supernatural creature, Mab, the Queen of Fairies. Bayanov and the people of Kabardia describe an almasty; is it a creature close to *Homo sapiens* or a being escaped from the world of spirits? Among many so-called traditional people, the distinction between the world of the living and that of spirits is not as sharp as one might imagine.

For now, let's look more closely at the work of Dr. Koffmann. We have already seen that she works, within the Russian sphere, with devoted collaborators ready to face the hardships of fieldwork and a harsh climate. That is truly the price to pay to approach and understand the local people and the fauna.

---

1   William Shakespeare, *Romeo and Juliet,* Act I, Scene IV.
2   Jean Froissart as quoted by Christian Le Noël in *La Race Oubliée* (tome 2), p. 216. Christian Le Noël adds: "This proves again that Wild Men were still numerous at that time and well known to the general public."
3   Dimitri Bayanov, *In the Footsteps of the Russian Snowman*, p. 62

21. Myth and Reality

"I am not the one who projected a myth onto reality; quite the contrary!"

So did Marie-Jeanne Koffmann exclaim at a conference in Paris in 1984.[1] Following her predecessors and collaborators, she delved deeply into the realm of mythology. It is a pleasure to be able to quote one of her few publications. While her oral presentations were models of clarity and literary quality, one can only deplore the paucity of her written works, equally rich and precise:

> In 1957, when Soviet investigators began, with great circumspection, to consider the problem of the so-called Snow Man, their first step was to look into mythology and to turn to the past: the absence of any trace of such beings in those domains would have put to rest any hypothesis as to their current existence. The initial results of their enquiry exceeded their expectations: everywhere, throughout history, the Wild Man, be it *Homo sylvestris* or *Homo troglodytes,* is found side by side to this other biped, *Homo sapiens,* feeding the superstitious fears of shepherds, the curiosity of naturalists, the embarrassment of theologians, and the ruminations of philosophers. It is present in all mythologies.[2]

Marie-Jeanne Koffmann selected remarkable examples. For example, a Phoenician cup describing a hunting scene, an artifact dating from the sixth century BC, found in the Kourion trove in Cyprus between 1865 and 1876. Prehistoric frescoes are well known for their realism, the wall paintings of Lascaux being one example among many others. The renderings of animals by the artists of antiquity follow that tradition and delight both the amateur and the professional zoologist. In this scene, the climax of the hunt focuses on the flight and subsequent capture of a hairy biped, eventually tied up and felled with an axe.

Historians remind us that during his 425 BC voyage of exploration, Carthaginian navigator Hanno had reached the Gulf of Guinea, as related in the account of his voyage, the *Periplus,* preserved in a Greek translation. Some specialists maintain, however, that Hanno did not reach further than the Canary Islands.

Nevertheless, Hanno mentions his encounter, on an island, with wild men that his interpreters called *Gorillaï*. The explorers captured and killed three females, bringing back their skins to Carthage. The term Gorillaï, first encountered in the Greek text, originally applies to hairy human beings before becoming the "gorilla" of zoological nomenclature thanks to the American, Thomas S. Savage (1847).[3]

A stylized bison, Lascaux cave painting. PHOTO: Author's file

Assuming that Hanno could not travel far enough to actually see gorillas—whose first authenticated skeleton was found in 1852—who were these hairy creatures? "If people of antiquity weren't aware of the existence of gorillas, did they know about the existence of large, surprisingly human-like large primates?"[4]

Next, Marie-Jeanne Koffmann draws on a 5000-year-old Sumerian literary masterpiece, the *Epic of Gilgamesh:*

> Enkidu, since you were humanised and left
> What man has done to the animals
> That trusted, and that drank with you
> At the desert pool!
>
> We are not beautiful as they are,
> We are not true.
> Our innocence and trust went with Enkidu
> When he deserted his friends for love of man.[5]

The Sumerians lived in the fifth millenium BC and were the inventors of cuneiform writing, drawn with reeds on soft clay tablets. Their brilliant civilization left rich traces on the people who followed them in what later became Persia (Iran–Iraq): Hammurabi's first code of law, the dazzling architecture of temples, the towers and ziggurats that crowned them. Not to forget that the Sumerians probably invented written account-keeping before the development of actual writing. From then on, transactions could be easily and faithfully recorded.[6] Equally elaborate was the mythology of the country that later became Babylonia and Assyria. Who could forget the moon goddess Ishtar, or Marduk, the bull god?

What really matters here, however, is the hero of an epic poem, Gilgamesh, a historical king from the third millenium BC, said to be one-third human and two-thirds divine. When his subjects complained to the gods about his despotic rule, the latter created, as a counterbalance to his authority, a wild man, the friend of the animals. The king told his mother about the dream that haunted his nights. She answered that the dream informed him that a companion was chosen for him, a brave comrade who used to help his friends in need. The man would come to him.

In his skillful interpretation of the epic, American writer Robert Silverberg relates the "fantastic and supernatural" adventures of Gilgamesh in a realistic fashion, so that the purely mythical elements stand out sharply. Enkidu, the wild man, is described as going naked, an enormous hirsute brute covered with coarse hair, closer to beast than to man, a savage creature of the wild. He behaved like an animal. He grunted, and squatting, ran easily and with agility on all fours on clenched fists and feet.[7]

Enkidu browses with the gazelles and drinks from the same springs as they do. He destroys the hunters' traps and nets. Marie-Jeanne Koffmann comments that he protects wild life, albeit "strangely."

King Gilgamesh knows the recipe that will tame this beastly creature. He orders a sacred prostitute, the very one that initiated him, to transform Enkidu and prevent him from reverting to savagery. In the embrace of a woman, he becomes like a man—this woman with her long lustrous hair who appears, behind the guise of the sacred courtesan, as a shadow of the future Mary Magdalen.

The courtesan leads Enkidu to the gate of Uruk and introduces him to the shepherds, who welcome him and offer him bread and wine. He leaps with joy! He is the messenger of the gods, awaited by the people, the one who will take a stand against the arrogant and unfair ruler. At a nobleman's wedding, Enkidu prevents the king from insisting on his "droit de cuissage," the Lord's right of first night with the bride.

Enkidu and Gilgamesh fight, but find themselves of equal strength. After a struggle as furious as that of raging bulls, the two heroes make peace. The king recognizes his double, the friend and brother announced by his dream. Enkidu abandons the company of shepherds in order to live with the king. The wild man becomes the king's right hand: Neanderthal, for a time, is to be *Homo sapiens'* closest companion and a model for all.

The transformation of the wild man is completed by the act of shaving, prescribed by the king who sent his barbers and surgeons to grind down and polish away the last traces of his savage past.

Thus emerges an early sign of a mellowing to appear much later, in the Age of Love, after the advent of the Ram, god of justice, meaning Yahveh. But for now, we are still in the Bronze Age, in Mesopotamia, where, under the sign of the bull, people are building cities, pyramids, weapons and jewels...

Gilgamesh is supposed to combat false gods and the heresy that is shaping up. However he is somewhat off-track: he was supposed to fight Enkidu, not to befriend him! As a king, he finds himself bound by his royal status. The wild man came to help him, giving him a brief taste of freedom and brotherhood.

In that marvelous article, Marie-Jeanne Koffmann guides the reader through to the first Chaldean Empire (about 1500 BCE). Quoting from the *Epic of Gilgamesh,* she strongly recommends it to the reader, encouraging him to discover a cyclical progression of eras—of the Bull (Taurus), the Ram (Aries) and the Fish (Pisces) leading to that of Aquarius. Enkidu, the wild man, has brought us to this view of Sumerian civilization, an unexpected development that nevertheless emerges implicitly from the adventures of the epic.

---

1   Annual meeting, of The International Society of Cryptozoology at the University of Paris VI. Many chapters of my previous work, *Sasquatch/Bigfoot and the Mystery of the Wild Man,* refer to Koffmann's work.

2. Marie-Jeanne Koffmann, article "Les Hominoïdes reliques de l'antiquité," *Archéologia,* no. 307, page 34. December 1994.
3. Savage T.S., Communication describing the external character and habits of a new species of Troglodytes (*T. gorilla*). Boston Soc Nat Hist: 245–247, 1847.
4. Marie-Jeanne Koffmann, op. cit., p. 38.
5. "Enkidu – a Defection," a poem by Odette Tchernine in *The Singing Dust,* Neville Spearman, London, 1976, p. 75.
6. See Georges Ifrah, *Histoire Universelle des Chiffres,* Seghers, Paris, 1981.
7. See Robert Silverberg, *Gilgamesh the King,* Arbor House Publishers, New York, 1984.

22. Dr. Koffmann's Conclusions

Koffmann's conclusions are based on those aspects of the Gilgamesh epic which provide clues as to Enkidu's real life: a child of the steppe, brought up by a gazelle, nursed with the milk of the wild ass, he is familiar with the forests and the mountains. Other signs expose the primitive features of this wild man, a mere sketch of a hominoid, member of "a population of anthropomorphic creatures." Far from being the king's brother, Enkidu was more likely to have been his slave, roped in to shepherd his flocks or perhaps used as a game tracker "helping to track down creatures similar to itself but perhaps even wilder and more beastly."[1]

Drawing from the Bible, Marie-Jeanne Koffmann quotes the words of Jehovah to Rebecca: "Two nations grow in your belly, two people will split apart after emerging from your womb; one will dominate the other; the first born will serve the younger." (Genesis 25, 22–24)

Koffmann interprets: "So, who is the first-born, which of the two people appeared first? A hirsute creature covered with red hair… spending all his time running in the wild…some kind of a brute, hardly able to express itself, unable even to give a name to the plate of lentils which he requests, whining, and who, driven by his brutal voracity, does not hesitate to trade in a birthright that he is clearly unaware of."[2]

The brute is, of course, Esau, whom younger brother Jacob tricked out of his birthright. The two brothers are finally reconciled 20 years later (Genesis 33:4). No one can fail to note the similarity with the wild Enkidu.

Marie-Jeanne Koffmann's strategy consists in extracting from the mythical plane details of the daily life of the wild man. That is indeed the goal of cryptozoologists:

> To bring mythified animals back to their normal proportions and especially to discover their true behavior, it is necessary to remove those aspects clearly associated with myth. This is usually rather easy, given the stereotypical character of myth. It then becomes possible to sketch a snapshot capturing both moral as well as physical traits, based on the descriptions by various eyewitnesses, a portrait which of course becomes more detailed and precise in proportion to the number of observations.[3]

There is an impressive array of valuable eyewitness reports—over 500 by 1991—arising from the people of the Caucasus. A number of prominent thinkers, such as, for example, Professor Piveteau, a member of the French Academy of Sciences and honorary professor at the College de France, were impressed:

> The general consistency of the reports gathered by Doctor Marie-Jeanne Koffmann, and the fact that they stem from people of different cultures speak strongly in favor of their veracity…Should the authenticity of the Caucasian hominid be definitively established, we will face a creature in which humanity and bestiality are intimately joined…This hominid raises the ever-recurring myth of the wild man which, as late as the Age of Enlightenment, was still pictured as a hirsute creature combining human and animal features at the vague limits of humanity.[4]

While I would love to quote at greater length from the Caucasian reports gathered by Dr. Koffmann, a few informative elements will have to suffice. For example, even a listing of the nutritional regimen of the almasty is fascinating:

- **Wild plants:** fruits and berries (oak, walnut, mountain ash, dogwood, wild roses, strawberries, blackberries, currants); shelf fungus growing on trees; green plants floating on ponds; chervil, hogweed, bugloss, capsella, lichens, sorrel, cohosh

- **Cultivated plants:** all species, without exception (garden plants, fruits, cereals, oleaginous, etc.)
- **Animals:** mice, rats, ground squirrels, squirrels, bats, newborn fowl, placenta of domesticated ungulates, carrion, horse droppings, frogs, frog eggs, lizards, turtles
- **Mineral foods:** rock salt, concretions along the edges of mineral springs (numerous in the Caucasus), clay, etc.
- **Foods taken directly from people:** milk, sour milk, cheese, eggs, bread, flour, bran, meat, honey, cooked dishes (soups, stews, etc.), jams

The almasty are so common that Marie-Jeanne Koffmann does not hesitate to state, "the relic hominoid has become a commensal of man."

An old shepherd familiar with the habits of the humanoid declared: "It eats all the wild plants that man used to eat when he was still wild."[5]

There is thus a close kinship between man and the almasty, if only in the realm of alimentation. Reports from local peasants are striking, especially in view of their precision and in the richness of details. Their reports are longer than modern accounts, which strive to keep information as terse as possible, restricted to a couple of sentences. Used to modern media—radio and television—the modern reader must not be bored!

Marie-Jeanne Koffmann's description of the eating habits of the almasty was criticized because some areas where almasty have been seen are lacking key elements of their food: they couldn't survive there. She answered these criticisms through an analysis of the best-known cases of wild children, normal children surviving in nature following special circumstances: lost in the forest, abandoned, kidnapped.

A famous case of wild children is that of the girls Kamala and Amala, the "wolf girls of Midnapore," aged seven to eight years and 18 months, apparently discovered in 1920 in a wolf's den in East Bengal.[6] According to the story of the children's rescue, the wolf protected them as fiercely as her own cubs; she had to be killed to rescue the girls, who appeared wilder than the cubs, and who scratched and bit the nuns who approached them. The children hated cooked food, preferring uncooked meat eaten off the floor, a preference countered by Rev. Singh, the rector of the orphanage that took in the girls, moving the food up. When, in 1921, Amala died of a kidney infection, the survivor, Kamala, turned to a hyena the nuns had been given to keep her company; however, soon she had to be separated from the hyena to humanize her. The Rev. Singh and his wife tried to civilize the little girl, who walked on all fours. They tied sticks to her legs to force her to walk upright. Kamala learned to do it, without abandoning her previous gait, when she wanted to move faster.

Marie-Jeanne Koffmann remarked that a dog would have been much better than a hyena. In fact, a child is what is made of him or her: when raised among sheep, a child "baahs" and is called a "sheep-child" (in Ireland c.1670); a bear-child in Lithuania, as described by Linnaeus in the mid 1700s as *Juvenis ursinus lituanis;* and even today in 2009 in Siberia, where five-year-old Natasha was found living with cats and dogs. She behaved like a dog, jumping and barking.

Wild children, whether really raised by animals or rendered wild by abusing adults, show remarkable physical strength and adaptability. After studying their lives, Marie-Jeanne Koffmann concluded that the almasty are even better prepared to face the elements, and they can be just as frugal and survive on a limited diet when forced by circumstances.

Koffmann found the case of Victor of Aveyron of particular interest because it clearly showed the results of the work of Dr. Jean Marc Gaspard Itard, a physician who spared no effort and showed a surprising degree of imagination in his struggle to humanize the wild child. When Victor walked out of the woods on January 9, 1800 near Saint-Cernin sur Rance, a small village of 350 people in France, he was 11 to 12 years old and walked upright, but everything else about him was animal. As such, he was not really different from the known cases (300 or so) of *Homines feri*. What distinguished him, the famous *enfant de l'Aveyron*, was the rapport that he developed with his mentor, Dr. Itard.[7]

When Victor was caught, he was in the act of digging out vegetables in a tanner's garden. He was offered a plate of meats, both cooked and raw, rye and wheat bread, pears, grapes, nuts, chestnuts, acorns, potatoes and an orange. He grabbed the potatoes, threw them into the fire and then picked them out of the embers without pain.

Such an instance of survival on a restricted diet illustrated the remarkable resilience of wild children, in support of Koffmann's views.

1 Marie-Jeanne Koffmann, *Archéologia*, no. 307, p. 39, Dec. 1994.
2 Marie-Jeanne Koffmann, op. cit., p. 40.
3 Bernard Heuvelmans, *Le Dossier des Hommes Sauvages et Velus d'Eurasie,* unpublished manuscript, 1997, pers. comm. to the author.
4 Communication from Jean Piveteau (June 10, 1987) to Marie-Jeanne Koffmann who inserted it in her article "L'Almasty, Yéti du Caucase," *Archéologia,* no. 269, June 1991, p. 37.
5 Marie-Jeanne Koffmann, "L'Almasty du Caucase – Mode de vie d'un humanoïde," *Archéologia,* no. 276, Feb. 1992, p. 57.
6 See Michael Newton, *Savage Girls and Wild Boys: A History of Feral Children.* Faber and Faber, 2002.
7 See Lucien Malson, "Les Enfants Sauvages," *éditions* 10–18, Paris, 2003. Also Roger Shattuck, *The Forbidden Experiment —The Story of the Wild Boy of Aveyron,* Kodanska International, New York, Tokyo, London, 1994.

23. The 1992 Almasty Expedition

Here we are in Paris in 1992; preparations for the Franco-Russian expedition are well underway. Movie director Sylvain Pallix is in charge of logistics as well as of filming a documentary. The film, *Almasty, Yéti du Caucase,* was aired on the France-3 television channel on February 13, 1993.

Dr. Koffmann is the scientific leader of the expedition. Although 73 years old, she is fit and spry, thanks to her background as an alpinist. Some of the members of the expedition's film crew, on the other hand, will have difficulty following the steep trails of the Caucasus, up and down ravines and nearly impassable gorges.

Marie-Jeanne Koffmann has been a regular visitor to a region stretching between the Black Sea and the Caspian, more specifically Kabardino–Balkaria, which she knows well. The almasty, of course, know nothing of human boundaries and roam over hundreds, even thousands, of kilometers. Today, the countries where they have been seen are troubled by violent conflicts: the Caucasus is in flames!

To this day, there is no photo of the almasty. Marie-Jeanne Koffmann explains to the journalists that Causasian peasants, just as those at home, do not burden themselves with cameras when they work in their fields. In the mountains of the Caucasus one finds mainly farmers and shepherds. Their life is hard, the standard of living modest and luxuries are unknown. A variety of electronic gadgets, common in richer countries, are a novelty for the locals. The Caucasian villagers are put off by the extravagance of supplies and equipment, and especially by the arrogant attitude of the film crew and of the various technicians. Marie-Jeanne Koffmann already complained about the media hubbub preceding their departure. She told a journalist, "For these simple and wonderfully welcoming people, the existence of the almasty is obvious. They have to be rewarded materially. I'm sure we will not need two months…"[1]

Koffmann, normally patient and careful, is optimistic. Is she perhaps impressed by the wealth of equipment made available, hoping to benefit from its deployment? Of course, she hopes to be able to rely, as she has always done, on the local people, who have long known her. Undoubtedly, they will willingly trade their stories for some reasonable remuneration. A ruble is a ruble everywhere, especially when times are tough.

According to local people, the almasty walks like a human being, but its gait is reminiscent of that of a wild beast. It scares horses, as were those of Patterson and Gimlin when they met a bigfoot (sasquatch) in California on October 20, 1967.

The peasants say that the almasty turns its head like a wolf, meaning stiffly, together with its short but massively muscled neck and shoulders. Like its cousin bigfoot, it emits a strong odor and can run as fast as 40 kilometers per hour (25 mph). Unlike its American cousin however, it does not hesitate to feed a fire abandoned by shepherds; but it never lights a fire on its own. Sometimes it steals food from people, even clothing, that it sometimes puts on.

Everyone in these mountains is familiar with the almasty's habits. Two witnesses speak of the bright, "yellowish red" eyes of the creature.[2] Sylvain Pallix's documentary includes short interviews with people on the village street, at their farms and at home.

A horseman declares that the almasty can make itself invisible. A certain Micha, keeper of a well, saw an almasty through the window and reached for a weapon, but could not move, as if hypnotized into immobility.

A man named Hossim says that the almasty never show up during the day. They are rather repulsive, but are similar to people. Hossim also confirms that the almasty turn their head stiffly. He actually saw one in a tree in his native village, razed on Stalin's orders in 1940, a date he has not forgotten. Before the war and the arrival of machines, the almasty used to approach the villages. The exile of the

Balkars, forced out of their homes by the dictator's madness, accelerated the dispersal of the almasty.

According to another, it is best never to utter the name of the diabolic creature so as not to incur Allah's wrath.

The interviews on film did not turn up anything that wasn't already known. Nevertheless, the flow of interviews leaves an agreement on facts that may be compared to observations from elsewhere (Tibet, Nepal, Mongolia...).

Back in Paris, Yves Coppens, Professor at the College de France, had discreetly encouraged Marie-Jeanne Koffmann.[3] She could leave on the expedition with as much confidence as one could possibly muster. She was afraid that by revealing the secrets of the almasty, she would contribute to the end of the mystery, especially if one of the relic hominids was captured.

Her fears were unfounded, but she winced at the behavior of some members of the expedition. She was shocked by their contempt of the local people, to the extent that she long hesitated before returning again to the Caucasus. She believed, wrongly, that her long-time friends would hold her responsible for that misbehavior.

For Koffmann, the expedition left a bitter taste. She rarely appears in Sylvain Pallix's documentary, which merely skims the surface of the almasty phenomenon and leaves the viewer quite unsatisfied. Too bad, for there are so few documentaries on the wild man!

Gasik, Marie Jeanne Koffmann's jeep.
PHOTO: Author's file

The moving picture, be it documentary or fiction, turns out to be particularly ill-adapted to a subject that demands ambition as well as intellectual honesty. In the domain of cryptozoology, which the scientific establishment, except for some rare exceptions, considers pseudoscience, the French achievements are mediocre and often even coarse. For example, in 2008 Jacques Mitsch completed a film, *Almasty, la dernière expedition,* for the Arte cultural channel. It's hard to imagine a clumsier comedy. Could it be an act of vengeance toward Marie-Jeanne Koffmann? One can only hope that it's actually an unfortunate coincidence. Mitsch filmed on location—in the Pyrenees! He represents the Caucasian peasants as mud-stomping rustics. The dance of the shamans is straight out of a summer camp theatricals. The scientific leader, eaten up by ambition, exudes domineering authority; her teammate is a wimp. Compared to that film even *La soupe aux choux* (1981) appears as a masterpiece of refinement![4] Let's thus grant to Marie-Jeanne Koffmann the important place that she well deserves and forget as soon as possible those who pretend to film her life from the wrong side of the lens.

Originally quite skeptical about her research, for it seemed to her almost absurd that there should exist a humanoid creature in the Caucasus, Koffmann devoted the greatest part of her life to proving the existence of this "anthropoid ape" related to the orangutan. She knew that there was still no formal proof of its existence, but the reasons for continuing to make it the object of scientific research kept piling up.

With her deep erudition, her study of the history and prehistory of the Caucasus, her knowledge of local languages, her ability to forge long-lasting links with local people, Marie-Jeanne Koffmann is a living memory. Anyone who has had the opportunity to hear her conferences will confirm it.

Even more convincing, for those who have read them, are the four articles she published in *Archéologia*. They reveal the author's method, where she first addresses mythology in general, harking back to the deepest antiquity, and then focusing through more specific and prosaic myths on the concrete features of the wild man's behavior.

A particularly interesting passage is that where, after a reminder of Enkidu, the hirsute hero, Koffmann describes the taming of a "savage" from the Caucasus. He learns to do heavy work and to look after the herd; he is resourceful and perceptive, as faithful as a dog, and knows the forest like the back of his hand. In another passage, a cultural education delegate of the Communist Party's regional committee speaks of an elderly Balkar couple that harbored a female almasty:

> She knew how to perform many tasks: carrying wood or water, leading the herd to the barn or gathering it for the night under a rocky ledge…The Party official quickly found the right words to describe her role…in a word, she brought a physical contribution to the fulfillment of summer plans.[5]

The behavior of humanoids varies somewhat from country to country and from time to time, as does that of animals, and is of some interest. What is however most important is the method Koffmann used: starting from myth, she ends up immersed in local daily life. The almasty is part of it, with the characteristics which he shares with his cousins, as described by Eric Shipton, Peter Byrne, Véra Frossard, Odette Tchernine, Reinhold Messner, Boris Porchnev… without omitting ancient writers:

> One of the most extraordinary aspects of the problem of relic humanoids is their presence through the centuries in the company of man, involved in his history, his cultures and his home life, as quiet, discreet and elusive creatures. Each era had its own interpretation of this troubling look-alike without ever succeeding in defining it precisely, and it is amusing to see the labored interpretation given by today's ethnologists of these many "wild men," usually consigned to the realm of the imagination.[6]

The search for the wild man has generated a surprising mixture of tales and factual accounts, akin to a cartoon where live actors also appear, thanks to editing techniques. The overlap is such as to make it difficult to realize which domain is dominant. Generally, the characters are normal people, but there are notable exceptions. And sometimes, one escapes the realm of the known fauna to slip towards the unknown.

Can an overview bring out a thread with a sufficiently strong scientific basis to support the existence of the wild man? Might that elusive thread refer to a stage beyond our ken, a shadow theater where we occasionally catch glimpses of meaning?

Let us now look further and try to learn more from China, where reports about the wild man—the yeren—arise from many areas.

In a previous work, I noted the presence of wild men on all continents. I gave, for China, the example of the tragic demise of a hirsute creature in Tibet. The eyewitness was Chinese, the commanding officer of the Chengdu area, adjacent to Tibet. This officer had twice seen a snowman while he was stationed in the Ngari province (northwest of Tibet). One day, a yeti tried to take away the gun of a soldier, who defended himself and killed his assailant. The rest of the detachment said that they had buried the yeti without preserving any part of it.

There is a plethora of short reports, vague and unverifiable. What can we conclude?

---

1. Pierre Berruer, "La science traque le Yéti du Caucase," *Ouest-France,* 28–29 mars 1992, p. 6.
2. In my earlier book, *Sasquatch/Bigfoot and the Mystery of the Wild Man,* p. 209 ff, I described Marie-Jeanne Koffmann's discussion of nocturnal vision among almasty.
3. Yves Coppens, one of the discoverers of Lucy, added later: *"It* [cryptozoology] *is a perfectly respectable science...It coexists very well with zoology. Each year, a number of creatures pass from the realm of cryptozoology to that of zoology...All there is to it is that once fully documented, they go from one list to another."* He also added: *"Always expect the unexpected!"* www.rhedae-magazine.com/, Sat. Oct 20, 2007.
4. *Soupe aux Choux,* a film by Jean Girault. Two buddy farmers are visited by aliens who like their domestic cabbage soup.
5. Marie-Jeanne Koffmann, "Les Hominoïdes Reliques dans l'Antiquité," *Archéologia,* no.308, Jan 1995, p. 61.
6. Marie-Jeanne Koffmann, pers. comm. March 22, 2000.

24. Grover Krantz' Enquiry

Fortunately, there have been other serious enquiries besides Dr. Koffmann's investigations.

Grover Krantz was a professor of physical anthropology at Washington State University, in Pullman, Washington. In spite of his reputation as a teacher, Krantz was widely criticized for the unfailing energy that he devoted to the study of the wild man of the Pacific Northwest, called bigfoot by white Americans and sasquatch by the Natives. Dr. Krantz published a solidly documented book, *Big Footprints, A Scientific Enquiry into the Reality of Sasquatch* (1992). It is one of the rare books written on the subject by a well-known scientist—a book also disparaged by some.

I heard Krantz speak on numerous occasions, often during formal conferences. I respect his work and I know that he followed a rigorous approach in all his investigations of the wild man. This is why the report of the expedition he joined in China (May–June 1995) to produce a documentary on the yeren deserves mention.

The expedition took place in the province of Guangxi, south of Sichuan and Hunan, facing the Gulf of Tonkin in the south and abutting on Vietnam in the southwest. It included, in addition to Grover Krantz, 10 Chinese, three scientists from Taiwan, and a Japanese TV crew. The objective was to interview two members of the Miao ethnic minority who had recently observed yeren.

About 15 Miao porters carried the supplies. The Taiwanese scientists also wanted to study the flora and the fauna as well as the human presence in the area. The group left Liu Zhou, 265 kilometers (165 miles) north of Nanning, the provincial capital, and drove

300 kilometers (186 miles) towards the northwest to reach, after six hours on the road, a small Miao village unspoiled by civilization.

The Miao (or Hmong) are a group of linguistically related native ethnic minorities of southwest China. They are distinguished by the colorful weavings and embroideries of their women's festive apparel. Women also wear magnificent silver jewels and sometimes striking tiaras,[1] decorated with stylized horns—a reminder perhaps of some ancient cult of Taurus the bull?

The cave paintings of Guangxi, the traditional songs, the animist practices related to shamanism, a panoply of medicinal herbs, and the art of embroidery are among the many features of interest to the curious visitor. The second Silk Road—actually the most ancient—started from Sichuan, crossed Yunnan and Burma (today, Myanmar) to reach India. Neighboring provinces (such as Guangxi) benefited from the trade route, a conduit of commercial, scientific, religious and cultural exchanges.

For three days Grover Krantz, then 63, climbed toward a high Miao village, walking along the low walls guarding the terraced rice paddies, ascending the stairs between the terraces and following a variety of steep trails. Big-nosed—as all Europeans appear to Asians—bearded and large, at 6 feet 2 inches and 210 pounds (1.9 meters/95 kilograms), Krantz was the butt of many jokes in which he was identified with the wild man.

Krantz noted that modernity had little impact on the life of the villagers. He was struck by how friendly and hospitable the Miao were, in spite of their rather modest means. A world unchanged for centuries, particularly regarding animist practices, greeted the expedition, which was led by Prof. Zhou Guoxing, paleoanthropologist with the Beijing Museum of Natural History and an old friend of Krantz. Zhou spent many years studying the problem of the yeren in China; his work is known internationally. The Sino-American author Paul Dong devoted a whole chapter to "Professor Zhou and Wildman Research" in his book *China's Major Mysteries* (1996).

Conversation with the Miao witnesses required the help of interpreters. Kranz spoke English to the Japanese translator, who translated into Mandarin for the local forester/policeman who, in turn, translated into Miao. The witnesses answered and the interpreters passed their answers back. The process was carried out seriously; the witnesses were impressed by the attention they were receiving and eager to do well. Here is a summary of the information obtained:

1. When standing up like a human, the yeren is two meters (6 feet 6 inches) tall.
2. Its legs are in the same proportion as those of humans.
3. Its arms appear longer.
4. Its shoulders are relatively broader.
5. Its face is also broad.
6. Its chin stands above its shoulders, a purely human trait.
7. Its nose is somewhat flattened.
8. Its forehead is high, as in humans.

The nearest yeren was three meters (10 feet) from the witness. He walked away much like a man.

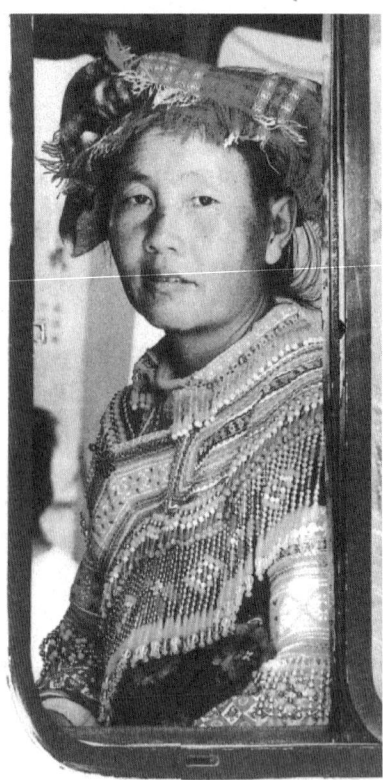

A Miao woman on a bus in Yunnan. PHOTO: Author

The second testimony described a squatting yeren, seen a few metres away, which stood up as he felt a presence, before running away using both hands and feet. Krantz thought that it had perhaps helped itself in climbing the hill by grabbing at bushes. His conclusion was that this south China yeren was different from the American sasquatch: shorter and less massive, with thick and long reddish hair, especially on its arms. These features were suggestive of a kinship with the orangutan rather than descending from *Gigantopithecus* (a species extinct for 300 thousand years).

An evolved orangutan, adapted to walking on its hind legs, is what the Miao witnesses described. However, the kind of vegetation available in such a craggy area would not suffice to feed a population of large apes. How could such a local variety of terrestrial orangutans feed itself, having adopted an erect posture and abandoned their arboreal habits?

Krantz thought that this was a new species, which he suggested could be called *Pongo erectus:* the standing orang. Alternately, it could be named *Yeren sinensis*, emphasizing its Chinese habitat.

Unfortunately, the 1995 expedition spent only 17 days in northwest Guangxi. The main problem was the difficulty of reaching the Miao village and the limited time available, circumstances outside the control of the reputable scientists who participated.

---

1   Some tiaras may weigh up to 10 kilos (22 lbs)! See Marie-Paule Raibaud, "Femmes d'une autre Chine," *Editions du Mont;* also picture of traditional Miao headdress at: www.travelblog.org/.

25. The Wild Man in Modern China

What was known about the Chinese yeti in the 1970s? Only a few people were aware of the testimonies gathered, especially in Hubei province. Such reports were nevertheless enough to justify the creation of a protected area in Shennongjia Forest, dubbed the "Yeti Reserve."

The local mountains are rich in legends. There was even a discovery of a 2000-year-old lantern depicting a *maoren* (hairy man) figure. In the fourth century BCE, a poet and statesman of the Chu kingdom spoke of the "ogre of the mountains." During the days of the Tang Dynasty (618–907), there is a reference to a group of "hairy men." Later, the poet Yuan Mei (1716–1798) describes a creature resembling an ape, without actually being one.

It was only in 1940 that the first scientific observation took place. Wang Zelin, a biologist and a graduate of Chicago's Northwestern University, had returned to China and was working for the Yellow River Irrigation Committee. In the fall of 1940 he was traveling by bus between Baoji and Tianshui. At the sound of gunshots, the bus stopped. A hunter had just killed a two-meter-tall (6 feet 6 inches) wild man. Its entire body was covered with long reddish-gray hair, at least three centimeters (1.25 inches) long. It was a buxom female, with a narrow face and sunken eyes. Her unruly hair was more than a foot long. The creature looked just like the plaster copy of the Peking Man (the Chinese equivalent of *Homo erectus*), except for a thicker and longer head of hair. With its thick protruding lips, the creature was extremely ugly. Local people said there was also a male around. The pair had been seen in the region over the past month. All they could utter were howls.

Why was this creature killed? The book that quotes Wang Zelin is silent on this matter.[1] During those warring years, many people with an itchy trigger finger were bearing arms. There, as in Russia, wild men were defenseless victims.

Let's now leap forward to 1976, a year marked by a portentous sighting. In mid-May a group of six managers of the Shennongjia Forest Service were traveling in their car when there appeared in the headlights a creature covered with reddish fur. It was neither a bear nor a known animal. The Beijing Academy of Sciences was immediately notified by telegram. The sighting created great public interest and prompted a sequel of military and scientific expeditions. Finally, the authorities set up an enquiry including scientists from Beijing and Shanghai and from the provinces of Hubei, Shanxi and Sichuan. In addition, the expedition included photographers as well as soldiers armed with sedative dart guns and accompanied by hunting dogs: altogether about a hundred people assisted by army scouts. Hundreds of people were interviewed between 1976 and 1977.

Together with local militiamen and commune members, the team

organized several large searches, but (as is usual with such expeditions) they found nothing definite.[2]

The Shennongjia is one of the rare remaining temperate virgin forests. The chalky soil is much eroded and harbors numerous dens and caves. The dense vegetation and the deep ground cover of rotting vegetation make progress difficult. The area is replete with specimens that botanists consider living fossils. Animals find easy refuge, for example, the takin, leopards, and the golden monkey, known locally as "snow monkey," found nowhere else and of which three varieties have been identified.

Myra Shackley writes: "My own view is that this unique flora and fauna provide the perfect refuge for an unknown primate, whether *Gigantopithecus* himself or his descendant."[3]

However, Suzanne Cachel, a professor of anthropology at Rutgers University strongly disagrees. In her review of Shackley's book, she wrote:

> Shackley is not the first person to connect *Gigantopithecus* with the Yeti or Sasquatch, but I emphasize that no proof whatsoever exists for this connection. There is no fossil or skeletal evidence detailing such an evolutionary progression…[4]

The issue remains the subject of debate among both specialists and amateurs.

Myra Shackley, admired for her clear, lively writing style, is also appreciated because of her typically British sense of humor. She suggests a "less scientific" hypothesis: In the days of the first Chinese emperor, Huang-Ti, laborers were forcibly drafted to build the Great Wall. Some of them hid in the forest where, after many generations, their descendants became wild and hairy men, although still capable of speech. From time to time, they would leave the safety of the woods and ask, "Is the Wall done?" and then, without waiting for an answer, they would flee back to the forest.

The Shennongjia golden (or snow) monkey is reminiscent of the Himalayan langur, also of medium size (up to 1.4 meters [4.5 feet] tall). Father Armand David, a French missionary and naturalist, discovered the Chinese langur in 1865. He identified the animal, which lives above 3000 meters (10,000 feet), from drawings on a piece of

Snow monkey. ILLUSTRATION: Alika Lindbergh

pottery. The snub-nosed ape was later given a zoological name: *Rhinopithecus roxellanae*.

With its thick fur and near-human nose it is easy to confuse the golden monkey with the snowman. However, the size and the shape of the latter's feet can hardly be mistaken for those of the golden monkey, a rare and threatened species, endangered by industrial forestry and forest fires. The plan to build a tourist airport near Shennongjia will certainly not improve the situation.

Unfortunately, the richness of the flora and fauna of this forest reserve seems to have been under appreciated. The importance of the life forms it shelters is to be measured by the importance given to it by scientists and cryptozoologists. For example, in 1982, Frank Poirier, a professor of anthropology at Ohio State University, went there in search of the wild man. He was accompanied by a pair of Chinese scientists: Hu Hongxing of Wuhan University, and Chung-Min Chen, another anthropologist from Ohio State. The reports they collected were similar to those gathered in China since the 1950s. The concrete items agree with previous investigations: tracks, hair, excrement.

Poirier also noted that they had found dens, the lowest at 1500 meters (4800 feet). In the summer, the wild man migrates towards the summits, which reach 3000 meters (10,000 feet) in that region. The wild man migrates in response to seasonal changes and fluctuations in the availability of food.

Animal bones were found in the dens, although analysis of the excrement suggested a mostly vegetarian diet, except for a few insects. In a natural shelter the wild man's nest is made up of seven or eight uprooted bamboo stems, which requires great strength. On top of that, the bamboo is arranged in the form of seats with a higher side: the back. These nests are very similar to those of the giant panda, which however have not been reported in the Shennongjia forest.

Poirier and his colleagues made surprising discoveries, linking geographically remote creatures from the USA and Canada (sasquatch/bigfoot), Nepal (the yeti) and China (the yeren). They noted: "What is most interesting is that such reports from these different regions have many striking similarities, despite the fact that, until rather recently, there has been minimal contact among the cultures of the peoples making the reports.[5]

In this respect, one notes that in spite of their political differences, China and the USA managed to collaborate at a scientific level in a domain that many "reasonable" people consider futile: cryptozoology. Poirier and his colleagues came to very conservative conclusions: concrete, irrefutable evidence remained insufficient to establish with certainty the existence of the Chinese wild man.

Poirier was just as circumspect as Prof. Zhou Guoxing, whom he mentions in numerous occasions in his publication. Zhou Guoxing had already visited the Shennongjia Forest before: he led the major 1977 expedition mentioned above. Some of the main results of his research are worth noting, first of all, the various names used in China for the yeren, an umbrella term covering various creatures: bears, orangutan, wild man, etc.:

- shangui: monster of the hills
- xing-xing: orangutan
- shandaren (in Fujian province): a man as tall as the hills, measuring more than three meters (10 feet).
- feifei (in some areas of Sichuan): a bear-man

These creatures are of such different statures that it is difficult to think that there might be just one kind of wild man. Zhou Guoxing notes that in Henan province, a wild man that was captured was skinned and its paws eaten as a delicacy: a gastronomical reminder of the plate of bear paws which has now officially disappeared from the menu of the best restaurants in China.

Many scientists seriously doubt that the yeren is anything but a bear or an ape. Many think that its presence is explained by hallucinations or hoaxes. However, Zhou Guoxing, as well as his colleague Wu Dingliang, an anthropologist from Fudan University in Shangai, thinks that the snowman is a tall, undiscovered primate. Zhou adds that it is the subject of numerous legends, not only in Tibet and Sinkiang, but also in the northwest parts of Yunnan, inhabited by descendants of Tibetans. Besides, there are close similarities between many of the witnesses' reports, suggesting the existence of real creatures. Zhou Guoxing believes that investigations by Chinese scientists in the thickly forested mountainous areas could reveal the existence of an unknown creature that must be described scientifically.

Even if the majority of Zhou Guoxing's colleagues reject the hypothesis of a humanlike being, others suggest that the yeren might be a descendant of *Australopithecus,* particularly of *Australopithecus robustus* or *Paranthopus,* the near-man, which lived two million years ago.

Chinese langur (*Rhinopithecus roxellanae*). PHOTO: Author's file

Still others believe it might be an ape, a descendant of the orang-utans, which were abundant in southern China during the Pleistocene between two million and 10,000 years ago. It could also be a descendant of *Gigantopithecus.*

The 1977 expedition collected casts of footprints, hair and excrement attributed to a creature resembling both a human and an ape. The study of the local fauna and flora, the corpus of mythological tales and legends, and the information gathered among the residents added up to an impressive body of indices. As for the almasty, the yeti or the sasquatch, those researchers who looked and listened without a priori bias reached interesting conclusions. For example, French zoologist François de Sarre:

> One might imagine that many human races survive on our planet under favorable climatic conditions, in spite of the current predominance of *Homo sapiens*. That has not always been the case, and the situation might change some day. In any case, there is no lack of reports of wild, hairy men throughout the world. These are studied and classified by hominologists, a term introduced by Russian researcher Dimitri Bayanov, as well as homins to denote those creatures, anatomically similar to man, but still in the wild. The best known of these, the yeti, is most likely an ape, a relative of the orangutan. However, there might perhaps remain a few *Homo erectus* in the Himalaya or in the Caucasus, even perhaps neanderthalians (Barmanu) in northeastern Pakistan.[6]

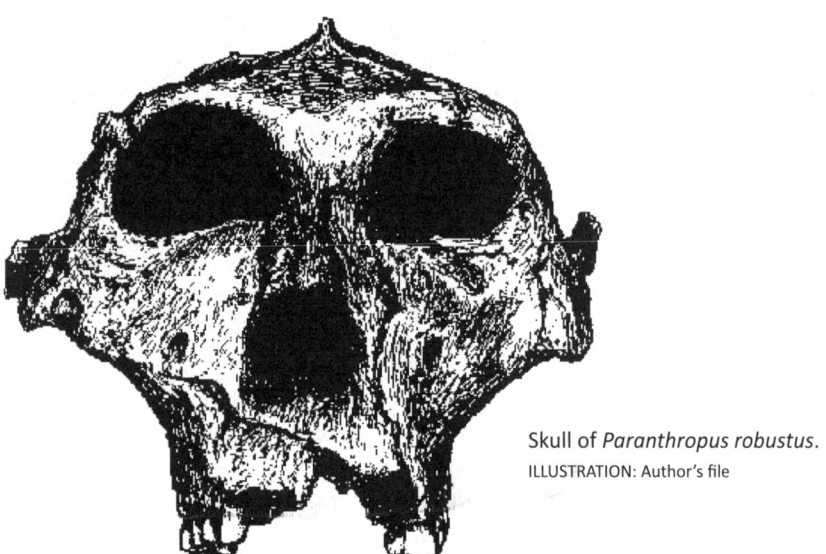

Skull of *Paranthropus robustus*.
ILLUSTRATION: Author's file

Zoologist Bernard Heuvelmans, who introduced the term cryptozoology, described three distinct types of wild hairy man:

- The small yeti, an ape denizen of the Siwalik mountains of India, a range parallel to the eastern Himalayas. That creature, the size of a human adolescent, has thick dark red fur, and a conical skull topped by a bony sagittal

crest, as found in gorillas and male orangutans. It is by all evidence an anthropoid ape, a pongid.[7]
- The great yeti lives further north, in Tibet, China, Burma and the Malay peninsula, roaming perhaps as far as the Himalaya. Because of its size, one tends to relate it to the fossil *Gigantopithecus*, whose remains indicate its presence between eight million years and 300,000 years ago. Reports often tell of the great yeti rather than the small yeti, as it is more impressive to have seen a giant. The fossil remains of *Gigantopithecus* have frequently been found together with those of orangutans and giant pandas, which have survived to this day. Why could a late representative of *Gigantopithecus* not have also survived?
- The third type of wild man is found from the Caucasus through Siberia all the way to the Malay peninsula. A specimen examined by Heuvelmans, which he named *Homo pongoides* was said to come from Vietnam.[8]

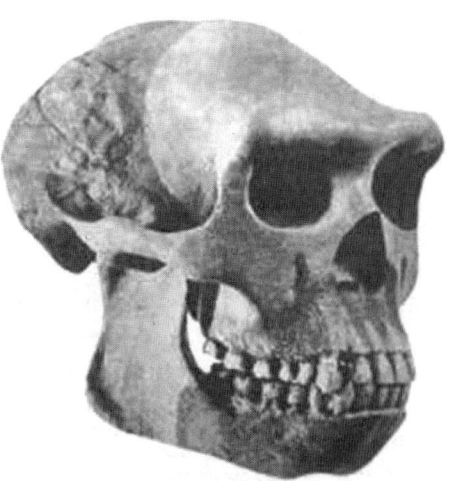

Skull of *Homo erectus*, Java.
PHOTO: Author's file

Clearly, the hypothesis put forward by Marie-Jeanne Koffmann and Bernard Heuvelmans that a great ape—either of the pongid (orangutan) family or perhaps a cousin of *Gigantopithecus*—might have survived is shared by many Chinese investigators.

One cannot simply brush aside these hairy creatures, anatomically different and behaviorally incompatible with any *Homo sapiens*. The list of reports keeps growing. New observations surface, sometimes from the same places. Twenty years after Professor Zhou Guoxing's 1977 expedition, an article in *China Daily* describes a creature half-man half-ape, dubbed "bigfoot." In September 1999,

Professor Zhou Guoxing.
PHOTO: Courtesy of Prof. Zhou Guoxing

a team of scientists, journalists and local worthies trekked deep into the Shennongjia Natural Reserve and found (40-centimeter [16-in]) tracks, chewed up corncobs, brown hair at the spot where a hunter had seen a strange "beast."[9] Yuan Zhenxin, a paleoanthropologist with the Chinese Academy of Sciences affirmed that the tracks were not those of a bear.

Professor Zhou Guoxing remains doubtful as to the existence of the wild man in China. However, even though he might reject 95 percent of the evidence, he holds it necessary to study the remaining 5 percent.

In China, as in the Caucasus, Tibet or Nepal, the professional explorer, faced with the mystery of the wild man, progresses at a snail's pace and must learn to be content with minimal clues. The giant of the forest and the mountains has become rarer, although less so than the vegetal species that cover the summits of the Shennongjia, shrouded in mystical mists.

Shennong, the divine farmer.
ILLUSTRATION: Author's file

Shennong is the name of a god-king. Shennong, the divine farmer, creator of agriculture, invented, among other things, the axe, the hoe and the plow. The word *jia* means "scaffold," which had to be built as a giant ladder, rising day by day to reach the summit of the mountain where grew the medicinal plants. After a year the subjects of God-King Shennong finally reached their goal. The king tasted all the plants and it is said that he discovered a tea that was the antidote to at least 70 toxic plants, which he tried on himself, all in the same day. As the father of Chinese medicine, he is also credited with the invention of acupuncture.

The forest harbors splendid specimens of tulip trees, rising up to 40 meters (130 feet). Magnificent dove trees, *Davidia involucrata*, named after Father Armand David, sport large white bracts which surround flowers looking like folded handkerchiefs. Groves of arrow bamboo, *Pseudosasa japonica,* five to six meters (16–20 feet) sometimes create impassable curtains, providing shelter to animals and a treat to the giant panda. The *metasequoia,* or Chinese sequoia, first identified in 1943, dates back to the Pleistocene. It is a living fossil, the only existing *metasequoia* reaching 50 to 60 meters (160–200 feet) in height.

Why is there so much interest in the trees of the Shennongjia? Because the wild man lives in nature, and preferably in the forest; it finds there its subsistence as well as the air it breathes. Its environment—the terrain, the fauna, flora and people that surround it—is mine and yours to discover, humble researchers that we all are. This discovery shall be the reward of our quest.

---

1 That book is *Still Living?* (1983) by Myra Shackley, British archaeologist and world traveler. The book speaks of the yeti, the sasquatch and the enigma of the Neanderthal.
2 Myra Shackley, *Still Living?* p.83
3 Myra Shackley, op. cit., p.90.
4 Suzanne Cachel, "Book Reviews" in *Cryptozoology,* vol.4, 1985, p.95.
5 Frank E. Poirier et al, "The Evidence for Wildman in Hubei Province, People's Republic of China," *Cryptozoology* vol.2, Winter 1983, p. 25.
6 François de Sarre, "La Théorie de la Bipédie Initiale," *Bipédia,* no. 26, Jan. 2008.
7 Bernard Heuvelmans, *Le Dossier des Hommes Sauvages et Velus d'Eurasie,* unpublished manuscript, 1997, pers. comm. to the author.
8 This creature truly deserves to be called a wild man, for Heuvelmans describes it as "undoubtedly hominid." For more details on this sub-species of *Homo neanderthalensis*, see *L'Homme de Néandertal est toujours vivant.*
9 "Big search for 'Bigfoot' under way." *China Daily,* Monday, December 6, 1996, p. 1.

26. Conclusion

Why is there such an interest in "monsters?" This question remains relevant today, in the age of science and transparency. The monster surfaces, settles in, ready to haunt us. We are ready to welcome it, eager to feel a shiver of fright. We are keen to see it, be it dinosaur, Loch Ness monster or yeti. The wild man has now morphed from abominable to friendly, thanks to *Tintin in Tibet* and a plethora of stories about bigfoot and other friendly giants. Friendly? Perhaps not so!

In the Pacific Northwest, there are a number of reports of aggressive behaviors by sasquatch/bigfoot: attacking hunting Natives, pelting white loggers with big rocks, kidnapping women. The yeti is quite rightly feared: it can bring down yaks. And the almasty is also guilty of sexual harassment, the females being not the least enthusiastic in this matter.

In their practice of violence, wild men seem to behave as proxies for the genus *Homo*. Some will suggest, in bitter irony, that this behavior is the proof of their close kinship with man, more precisely with our brutish prehistoric grandfather, *Homo sapiens*. It is as if, while humanity becomes more refined, so would its more violent traits. That seems to be the theme of many a cave painting:

> The whole cave was a bone-yard, a gallery of animals in flight, so realistically depicted, so full of life, and so full of the horrible certainty of dying.[1]

Such is the vision offered by the spellbinding images found in the caves of the Dordogne, in France, or on the island of Levkas, in

Greece. It is on that island in the Ionian Sea that novelist Hammond Innes sets the fieldwork of his hero, an anthropologist and a genius, but at the same time a charlatan. The researcher has discovered the genesis of human nature: an inborn aggression, enhanced by a sense of organization, which has allowed Cro-Magnon man to eliminate its Neanderthal competitor.

After reading that excellent novel, more than 30 years ago, I remained stunned; the author is most convincing and draws on scientific results (Leakey, Dart…). I was still young and idealistic; that novel shook me up. Since then, I have refused to accept the idyllic vision of the wild man as a kind and benevolent being.

The impact of *Homo sapiens* on his environment should be viewed as honestly as possible. His adaptive capacity has made him a fearsome predator and conqueror. To survive, the wild man must remain well hidden. Serious people, especially scientists, find its presence disturbing. It is too much of a hybrid for many. Some even pretend that it breeds by inseminating human beings! Its mode of locomotion is ill defined, somewhere between biped and quadruped. What does it do? Does it live in groups? Probably not.

It shows up so rarely that it is impossible to imagine its daily habits. Do they exhibit some cyclic regularity? It must be affected by the rhythm of the seasons, the changes in temperature and food availability. What about the seasonal patterns described by Peter Byrne in Oregon?

In the Caucasus, the almasty/wild man has developed neighborly relations with peasants. Such cohabitation is rare, but it is equally unlikely and difficult with other great apes, or with bears. Impulsively we sometimes wish closer contact with wilderness, only soon to reject it: "It is the bane of wild people to wish to live like civilized people, and similarly that of the civilized to wish to live in wilderness."[2] To wish to live in closer harmony with nature, so often cast aside, is nevertheless a legitimate goal.

The wild man, a paragon of discretion, stands as a role model. In spite of its size, it treads lightly on the Earth, consuming only the food it needs. Even its footprints are becoming rarer.[3] To tread lightly is also the philosophy of Native Americans, which my friend Ed Fusch, known as "Prospector Ed," summarizes as follows: "Walk softly on the sands of time."

The wild man is close to the sorcerer, a creature of the night and of the shaman who communes with the spirits to cure the sick. Through their costumes, these figures merge with the hirsute giants. We have noted the role of the shaman and of the volunteer who dresses as a wild man. We saw that people hesitate to approach him, as he becomes, if only for a moment, the monster responsible for all the ills of the village and must be expelled, at least symbolically.

In chapter 4, I told the story of the tree with two branches: one branch bearing delicious fruits, the other poisonous ones; how one day a sick villager picked a fruit which cured him, how the other villagers rushed to cut the other branch and how the tree died as a result.

The wild man is part light, part shadow. When we look for it, are we not also seeking our own reflection, misplaced perhaps, if not entirely lost, as for vampires? Who would agree to lose his soul, if not through a pact with the Devil, out of despair, perversion or bravado?

The wild man might be our invisible self, reappearing from time to time, undeniable and belonging to a realm of ever-present giants… or is it perhaps a divine messenger, a forerunner?

While some seek enlightenment honorably, through initiatory practices, other plunge into the deep shadows of the forest where they hope to meet their fate under the shelter of the tree—some will see here an allusion to the Tree of Evolution.

I have rediscovered some notes scribbled a dozen years ago about the Neanderthal, sometimes identified with the wild man. At that time, I wrote, "these unassuming creatures, free of nuances, fascinate us by their simplicity and their guileless behavior. They are examples of humanity under development, evolving towards sapiens, still without refinement, complication, or duplicity. Some claim that in appearance, dressed as we are today, Neanderthals would go unnoticed on the street.

Speaking of duplicity, would those creatures be capable of guile or simulation? Would they breathe the fresh innocence of childhood? In any case, primitive creatures are incapable of the self-transformations performed by actors."

Might Neanderthal man be sufficiently evolved to pretend and calculate? If so, to what degree? It is certainly industrious and practical, but without pretense. Absolute and unrefined, it stands close to the wild man.

1. Hammond Innes, *Levkas Man,* Alfred A. Knopf Publisher, New York, 1971, p. 286.
2. René Laurenceau, *Les Anges et les Démons,* p. 11.
3. And yet, when I look at these plaster casts that I have from Mt. St Helens in Washington, I find that these feet look a lot like a variant of those of *Homo sapiens* or *neanderthalensis*. A large, robust and heavy creature!

Postscript

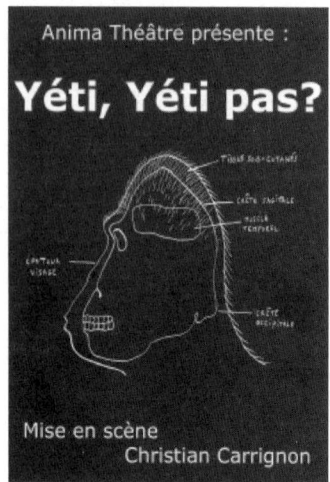

Poster for the play, *Yeti, Yeti pas?*
PHOTO: Anima Theater

Mise en scène
Christian Carrignon

I couldn't possibly finish without alluding to the show *Yeti, Yeti Pas?* ("Is it or isn't it there?" a simplified phonetic abbreviation of *Y est-il ou n'y est-il pas*), written in 2005 by two actor/puppet masters, Georgios Karakantzas and my nephew Gilles Debenat.

In a brief explanatory text, the director, Christian Carrignon, wrote: "These different interpretations of the "Yeh-the"—Tibetan for "the unknown animal of the crags"—are like shards of broken glass which, when put back together, create a mirror in which we can see our own reflection. For it is through the manner in which we look at others that we discover ourselves."

As a spectator, I was already sold. Nevertheless, I wish to emphasize the literary, poetic and humoristic qualities of this imaginary spectacle/novel/voyage.

What came to mind then were the words of Bernard Heuvelmans, relishing in advance his forthcoming, and last, major voyage:

> As for me, next year I shall be tracking the *Gigantopithecus* in Western Malaysia, Burma (now Myanmar) and in the Shennongjia hills in China; the Orang pendek in Sumatra; the Batutut in Borneo; and perhaps a Neanderthalian just about everywhere over there (although, as far as I am concerned, this issue is closed and no longer a topic of cryptozoological interest)."
>
> — Bernard Heuvelmans, pers. comm, April 1992.

Appendix 1 From the Roof of the World to the Mesas of Arizona

1. Merely random encounters?

In 2007, I attended the 19th Montaigu Spring Book Fair, in Vendée, France. Among the many exhibitors, I picked the booth where journalist and author Gilles Van Grasdorff, an expert on Tibet, was signing his books. I bought his *L'Attrapeur de Pluie* (The Rain Catcher) "a long journey through the world of the Hopi and the Tibetans, where Life never stops," as the author wrote when signing my book.

I then spoke to him of my work on the wild man in North America and of my forthcoming book on the wild man in Asia. We spoke only briefly as people were lining up to have their books signed. Fortunately, "the Rain Catcher" kindly answered some of my questions.

His book is the story of an encounter, one that may perhaps recur. In the beginning, two "worlds meet through two exceptional men, the sixteenth Karmapa and the Hopi chief of the Shungopovi Bear clan."[1]

The encounter occured in 1974 when the sixteenth Karmapa (a Buddhist spiritual leader, like the Dalai Lama) fulfilled a prophecy attributed to Padmasambava, the founder of the most ancient lineage of Tibetan Buddhism, in the eighth century. That philosopher, theologian and magician was a medicine man capable of healing the darkest ways of human consciousness. He was also a prophet:

> When the fire bird flies and the horse runs
> On the roads,
> The Tibetan people will be dispersed
> Like ants on the face of the Earth,
> And Dharma will come to the land of the red people.[2]

The age of airplanes (civilian as well as military) of automobiles and tanks, of exile for many Tibetans, has unfortunately come, and the sixteenth Karmapa feels that it is time for the prophecy to be fulfilled in its entirety. The Dalai Lama himself, supreme spiritual leader—although not ruler—of the four Tibetan Buddhist lineages had to flee his country during the 1959 Chinese invasion.

During a trip to Arizona, the sixteenth Karmapa, known to Tibetans as the Buddha of Medicine, demanded to visit the Hopi Indians. Why? And why in particular the village of Shungopovi?

The Hopi, an ancient people, are thought to be the descendants of the Anasazi. After the First World, that of the Creator when the Clan of Fire was preeminent, came the Second World, that of the Clan of the Spider: all creatures understood each other and were short of nothing; there were no diseases. Then came the Third World—that of the Clan of the Bow—followed by the Fourth World where the Clan of the Bear dominates.

As with many other people, Hopi history unfolds in cycles. Among them, the seeds of the Fifth World have already been sown: one must learn to recognize them. The Hopi and the Navajo share many of the same problems. Cancer cases are proliferating. The government stocks nuclear waste in the sands of the surrounding desert. Chemical and petrochemical industries poison the air.

In Tibet, discharge sites are at the surface. Pollution in the water and the soil foster diseases, congenital malformations in people and beasts, and mysterious deaths. "Neither the Tibetans nor the Hopi gather any benefits from the frantic industrialization of their territory."[3]

When the eminent Tibetan visitor reaches the Hopi village, the area is still in the throes of a long and trying drought. In the car driving him there, the air conditioning breaks down. Although the temperature reaches 45°C, the traveler controls his metabolism and does not sweat. In the winter, he practices tuomo, the mystical inner warming. He is aware of the parallel prophecy among the Hopi, announcing the arrival from the East of a man wearing red. The Tibetan wise man, Rangjung Rigpé Dordjé, stops some distance from the village and walks unhesitatingly towards Ned, Chief of the village as well as of the Clan of the Bear.

The Karmapa tells him, "We are related and we are both enduring terrible pains. The time has come to speak to each other."[4]

Onlookers press closer, eager to witness this extraordinary encounter. Chief Ned, the Karmapa and other clan leaders enter the *kiva*, the inner sanctum of Hopi culture. The Fourth World is represented by the lower part of the sanctuary, where the altar is positioned. "Among the Hopi, the altar is an image of the world and its various stages. It corresponds point by point to the Tibetans' mandala."[5]

When the participants finish their chants and prayers to the spirits—the kachinas—clouds gather and soon, as the Karmapa returns to the valley, it starts to rain. The Hopi and the Tibetan wise man prayed together for the first time and "on that day of October 1974, they caught the same rain."

Some 30 years later in 2004, Gilles Van Grasdorff, an admirer of Tibetan spirituality and medicine, followed in the Karmapa's footsteps. Van Grasdorff had already written a biography of the Dalai Lama's personal physician, Tenzin Choedrack. Tibetan medicine resembles that of the Hopi in that it is holistic, treating not only the failing organ but the whole person. By chance, Van Grasdorff met a Hopi physician who guided him to the village of Shungopovi.

On March 20, 2004, the Hopi were celebrating the spring equinox. The journalist was invited to the night dance of the kachinas. The Abbé Pierre[6] used to say "life was a tent for the night." The night of the kachinas transcends the canvas of the tent and opens the door to the illusion, which is true reality. During that night, time ceases to exist. The kachinas, spirits rather than divinities, led Van Grassdorf to that place where the Karmapa and Ned prayed in 1974, for there is so much to do in these last days of the Fourth World.

The kachinas sing and dance for hours. Ned, the Chief of the Bear Clan, throws a kachina doll to the journalist. I leave it to the reader to imagine the spiritual enrichment associated with this inner journey! Such experiences are seldom to be measured with scientific instruments.

One should understand that kachinas, though possessing a single nature, have a variety of forms. Van Grasdorff compares them to the Tibetan dakinis, "those that travel through space." He also compares them to Christian angels or to medieval fairies. These feminine deities are usually wise and protective, although some might also be evil and terrifying.

To sum up, kachinas are intermediaries between the world of men and that of spirits, or of the masked dancers embodying these spirits, or even carefully decorated dolls. One might say that kachinas are the spirits of the departed. When a dancer wears a mask, he becomes the kachina, the spirit that the mask represents. The dancer moves to the songs, shaking a rattle made of a gourd attached to a stick. The kachinas dance the coming of spring, a basic celebration, the beginning of a new cycle.

Thus, history never stops, life continues in a permanent cycle, except if one is reborn as a kachina among the Hopi, or one reaches Enlightenment among Tibetan Buddhists.[7]

2. The kachinas and Joyce Kearney

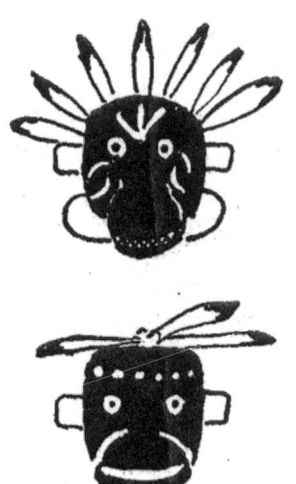

A pair of kachina ogres.
ILUSTRATION: Author's file

In my preceding book, an essay on the wild man of the Pacific Northwest, I examined the role of sasquatch/bigfoot among Native Americans. I wrote about the various aspects of Dzonokwa, giant female figure of the pantheon of the tribes of Washington, Oregon, Idaho and of British Columbia. Those tribes speak variants of the Salish language, comprising two main linguistic divisions: Coast Salish and Interior Salish. It is to be expected that Dzonokwa would appear under somewhat different names among different tribes.

However, beyond the domain of the Pacific Northwest, reports of the bigfoot emanate from the whole of North America as far as Florida where it is known as the skunk ape. There are even reports from South America: the *shiru* in Colombia, *guayazi* in Guyana, *oucoumar* on the Argentinean side of the Andes up to 4800 meters (15,000 feet), *mapinguary* in Brazil, etc.

Actually, the wild man is found on all continents. Author and illustrator Philippe Coudray, as nimble with his pen as with his brush, has inventoried the many types of bipedal hominids and drawn them after the testimonies of witnesses in a series of commented illustrations. "This book, which of course includes a certain amount of interpretation, has no pretension of being the result of a scientific enquiry.

It merely describes, as faithfully as possible, what witnesses, mostly natives, claim to have seen with their own eyes."[8]

Some 15 years ago, when the habitat of sasquatch/bigfoot seemed to me restricted to the Pacific Northwest, I was already convinced that, at least at a mythological level, bigfoot was everywhere. I discovered that eyewitness reports actually originated from all American states.

The small yeti.
ILLUSTRATION: Philippe Coudray

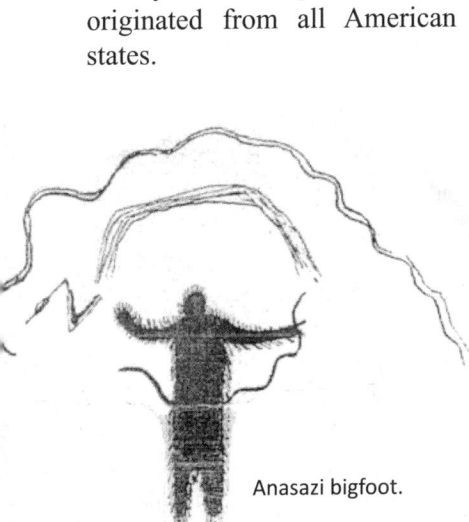

Anasazi bigfoot.

In January 2004, the cover of the monthly bulletin of the Western Bigfoot Society (*The Track Record,* no. 133) showed a pictogram from a cave inhabited by the Anasazis, the ancestors of the Hopi and the Zuni. This mysterious early American people had already disappeared before the arrival of the Europeans. They are thought to have reached their apogee sometime between 1000 and 1300 CE. Amateurs of history are fascinated by the quality of their weaving and artifacts, the skills of their farmers inspired by their god Kokopelli, their knowledge of astronomy, and their high level of urbanization. And to think these people knew neither writing nor the wheel.

Historian Joyce Kearney added a short commentary, noting that the pictogram represented a bigfoot leaving a cave near the Puerco River. She wondered about the nature of the object (a rope?) that it held in its hands. As to the double wavy squiggle above the creature's head, I wonder if it might be a sign of its aura, or of the spirit that guides every being.

In 2007, *The Track Record* no. 174 tickled the curiosity of readers by showing a hairy kachina doll, a possible bigfoot image. A photo shows a picture of Joyce next to the kachina Chaveyo and a mask of Dzonokwa, both objects belonging to her.

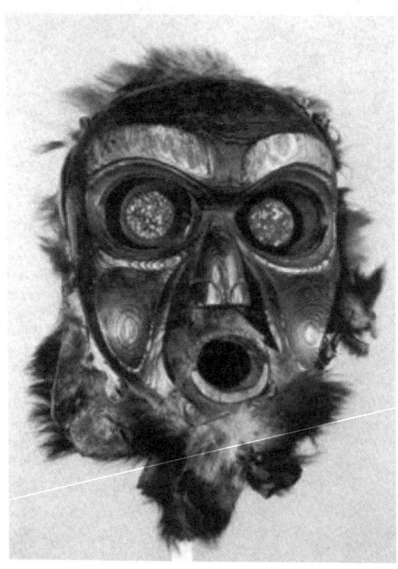

Dzonokwa, a Kwakiutl mask.
PHOTO: Christopher L. Murphy

The existence of a giant in southwestern North America is not an isolated case. The Paiute, the Hopis' northerly neighbors, spoke of Nu'numic, who lived with his wife in a vast cave from which he emerged to hunt the Indians, catch them, cook them and eat them.

Joyce's article also drew my attention to the tracings found near the Paiute Reserve in Mono County (eastern California). Along the Owens River a nearly inaccessible cliff, 800 meters (2700 feet) long, is covered with figures engraved in the stone (petroglyphs) and a few painted images (pictograms). These figures, the so-called Chalfant Valley Group, are only a few kilometers away from a series of tracks, a path seemingly leading to nowhere. There are hundreds of footprints carefully carved in the rock; some of them are those of giants. Were they indeed carved or are they real impressions?

Equally surprising are the ancient stone houses overlooking the Owens River. There are 150 of them on a deserted plateau, as well as villages of 40–45 houses such as at Fish Springs. These houses are round—four meters (13 feet) across and two-thirds to one meter (2–3 feet) high—sometimes far apart, sometimes close to each other.

They are built, without cement, of calcareous tufa, a stone commonly found in Mono Lake where it forms concretions that rise like towers above the water.

Joyce Kearney thinks that these houses, built before the arrival of the Paiute, were the work of the wild man, which she believes to be "more advanced socially and technologically than most people believe"[9]

These stone shelters are very similar to *bories*, dry stone huts found in France, used as temporary shelters for shepherds and their animals. The word *borie* comes from the Latin *bovaria* (a stable), a place for *bos/bovis,* Latin for bull. However, the wild men did not own cattle. Joyce Kearney concludes:

1. Men lived in groups, or assembled in groups, for the winter season.
2. They built permanent shelters in stone.
3. They used stone tools. How and what for?

Joyce adds that stone tools and arrowheads have been found in these villages, as well as the bones of marmots and mountain sheep.

While traveling in Colorado, Joyce asked the owner of a Native art outlet if he had a kachina representing Chaveyo. She was shown a black, very hairy ogre. She bought it. Kachinas are often bought today as decorative items or collection pieces.

Chaveyo's face is like that of an ape. It is an ogre who can emerge at any time in the spring to kidnap unruly Hopi children, reminiscent of the bogeyman of the Pacific Northwest tribes. Among the Hopi, other ogres, male and female—similar to the Basket Woman of the Makah Idians of Washington State—are close to Chaveyo: Soyoko and Soyok Mana (ogresses), Nata-aska (black ogre), Wiharu (white ogre) etc.[10]

Researchers who have studied the Hopi religion and those of their neighbors, the Zuni, Navajo and Pima, are astounded by the number of kachinas inhabiting their pantheon. Their religion incorporates deities from neighboring tribes and each clan has its own variant of the Hopi mythological system. It is easy to get lost among the crowd of kachinas. One may readily think that some are quite irrelevant, just like the bogeyman of our childhood. That would be a mistake. There are many categories or levels of kachinas, a word interpreted as meaning "esteemed high-rank initiates." For example, it is thought that the first category concerns the continuity of life. The kachinas of mid-winter dances herald the return of life: reincarnation is inherent to the continuity of life.

The second category belongs to the masters, those who teach us where to live, who we are and what rules our lives. The third category represents the "keepers of the law," who keep warning us until the day when they grow weary…

The Hopi universe is ruled by cycles, from the smallest to the greatest, resembling in this respect many other North American native civilizations. "Why is it more difficult to span 6 million years, a glaciation cycle, than 10 or 20 years for a solar cycle, than a few days?"[11]

Why should it be a surprise to find a religious leader, the Karmapa, followed by a Tibetan scholar, Van Grasdorff, make contact with the Hopi, another highly spiritual and threatened civilization?

The painstaking investigations of Joyce Kearney have revealed the breadth of the wild man myth. We now can appreciate better its rich complexity. If the bigfoot kachina were to correspond to a real flesh and blood entity, Kearney's suggestions would indicate an elaborate lifestyle: stone shelters, processional footsteps, cliffside petroglyphs.

The wall paintings of Lascaux and neighboring caves have surprised many observers. They were even thought to be a hoax; the art was too sophisticated, worthy of a Picasso or a Dali. Animals appear on top of each other, often doubled: mammoths, bisons, reindeer, horses…not to forget the surprising and superb shamans, half-man/half-beast of the Grotte des Trois Freres, in Aveyron. Who leads there? Who inspires the other, the man or the beast?

---

1 Gilles Van Grasdorff, *L'Attrapeur de pluie*, p. 11.
2 Gilles Van Grasdorff, op. cit., p. 35.
3 Gilles Van Grasdorff, op. cit., p. 179.
4 Gilles Van Grasdorff, op. cit., p. 131.
5 Gilles Van Grasdorff, op. cit., p. 135. A mandala represents the world, the cosmos, the palace of a deity, or, according to Carl Jung, a symbolic representation of the psyche. As a support for meditation, a mandala is

painted, sculpted in three dimensions, or crafted out of colored sand. There is some resemblance to the sand paintings of the Navajo, the Hopi's neighbors.
6   L'abbé Pierre: French priest who ministered to the workers.
7   Gilles Van Grasdorff, op. cit, p. 288.
8   Philippe Coudray, *Guide des animaux cachés,* éditions du Mont, 2009, p. 7.
9   David E. Stuart, *Anasazi America: Seventeen Centuries on the Road from Center Place,* University of New Mexico Press, Albuquerque, NM, USA, 2000.
10 Joyce Kearney, pers.comm. 2 July 2008. "Sasquatch being more advanced socially and technologically than most people believe."
11 Soyok litterally means "monster." I described the bogeyman role played by sasquatch/bigfoot in the last part of my book, *Sasquatch/Bigfoot and the Mystery of the Wild Man.*

Appendix 2
Angels and Demons

René Laurenceau taught Russian from 1966 to 1991 at the School of Mines in Saint-Etienne. I think of him as a prominent expert on wild men, from John the Baptist of the Bible to Bernard Heuvelmans' pongoid man. He has published in *Bipedia* (an online magazine) and in the monthly bulletin *Hominologie et Cryptozoologie*.

René Laurenceau is also a keen amateur of puppets. In Russia, these are often inspired by creatures from the folkloric tradition: russalka, tchutchuna or domovoi.

Laurenceau gave me a copy of a 40-page booklet entitled *Les anges et les démons* (Angels and Demons), which he illustrated with his own drawings of puppets. (The legend of the last sketch reads: "The puppet does not speak, but the voice speaks through it.")

The author has allowed me to quote from a few pages of his text. I love its poetic tone, lively style, insight and erudition. It opens up perspectives at once broad and deep on the question of the wild man.

Some preliminary definitions will be useful:

Russalka: a dangerous female creature living in the water, it may try to drown bathers. She is particularly active at the beginning of June (Russalka week). Anton Dvorák wrote an opera, in Czech, about this aquatic succubus, *Rusalka*, a lyrical tale in three acts (1901). Turgenev's tale "Pre Bejine" (in his *Stories of a Hunter*) features russalka as the heroine. It is also from Turgenev that Maupassant obtained the story "The Fear," where russalka attacks a young bather.

The Tchutchuna: a wild man with black skin and hair, from Yakutia in northwestern Siberia. Myra Shackley thinks he is extinct. René Laurenceau comments, "It is a Siberian belief that once a Yakut is near death, he must not be saved so that death will not be deprived of its promised. Formerly, a man rescued from the doors of death was expelled from all yurts and igloos. He had to wander in the tundra, chased away by all if he didn't want to leave. The belief was then that his skin would turn from yellow to black...The Yakut who cheated death was punished by becoming the tchutchuna of the tribe, a word derived from some primitive shamanism."

The Domovoi: "The domovoi has the stature of a child but the face of an old man. His blond, nearly white hair falls on his shoulders in wavy curls. The domovoi, whose name indicates that he is part of the household, lives under the stove...The survivor from the last glaciation doesn't fear the cold, but welcomes the heat from the stove... The domovoi visits the stable at night, pets the horses, tours the hen house to chase away the fox or the snake."

For additional details, please consult the article "Nègres blancs" (White Negroes), which Laurenceau published in *Bipedia* (no. 22, January 2004).

Excerpts from *Angels and Demons*

Chapter 8

According to legend, russalka has green skin and hair and wears a green dress. She is green all over, without worrying too much about the fact that demons need no dress. Which is not the case however for the angels Blue Beard, Red Beard, Black Beard and White Beard, who do need robes. The russalka needs no robe. Neither do the black Siberian tchutchuna, the golden haired Dionysus, the wild man of the woods from Java formerly called orangutan until the day that the mawas of Borneo stole its name, the naturally ochre-colored man of the Caucasus, quite different from the pale one which covered itself with ochre to hide from hunters. But the all-green russalka is but a legend. She covers herself with water plants to make believe she is green, just as the Neanderthal man calls himself neanderthal to suggest that he comes from the valley of the new man, when he is really an ancient man, returned like a ghost to the valley of the new man in 1856, the year of Freud's birth. We must be wary of the pranks of the wild men returning to us. Our spooks.

Chapter 27

Of the two anatomies, ours and the other's, the other resurfaces in 1856. It is the Neanderthal man, who lived before us. He was nocturnal. His ears were pointed, but not always. Nocturnal man on our planet shows great variety: in its ears, in the color of its skin: black skin, black hair, white skin, black hair, white skin, white hair; tanned skin, auburn hair. Nocturnal man has deep, round eye sockets, but no forehead. Everything is in the occiput. He is the first Adam. First Adam does not talk, but is comfortable in nature, his paradise. Second Adam shows up. That's us. Second Adam speaks. His children are not born easily. He finds caves uncomfortable. Hunting is too risky. Second Adam clears the forest and starts agriculture. He has to kill the herbivores. The carnivores also have to leave. First Adam, a hunter but not a farmer, also leaves. Second Adam worries. Who killed first Adam? Second Adam feels expelled from Eden. In the end, though, both Adams are but one. The Bible was right.

Chapter 28

In the theatre, a detail stands out: the nature of the pelt of the first Adam, pulled out of the closet. The robe is white, in a white cloth that doesn't fray; it is hemmed everywhere to prevent fraying. However, the hair is black, a black unhemmed and fraying material that makes the wild man look like a miserable tramp. Is the man of the woods just an unfinished man? Backstage in the theater, there is usually a store of hammers and nails, saws, ropes, paper, glue and cans of paint. The theater is obviously a construction. In the barn, one would like to find somewhere thread and needles, especially black thread to hem each one of the ribbons of the pelt of the Dionysiac creature. The Apollonian on the other hand, who speaks and sings in the Attic language, is hairless under his robe. The Dionysiac is as hairy as Esau. What a job, removing each hair and hemming it, to complete the costume! For lack of proper finish, the end looms. It seems that the whole character will unravel. The tchutchuna, the vodianoi, the russalka and the domovoi have disappeared and I challenge anyone to find in our world even a single Dionysiac.

Chapter 29

The cloth may fray, but the skeleton remains. We have not found the pelt of the paleanthrope, nor its nictating membrane, but his skeleton was discovered in La Chapelle aux Saints. And who knows? Perhaps we shall find complete Neanderthal. The frigid Siberian ground may one day yield a frozen man identifiable from its black skin as a tchutchuna. With a little luck, one might have succeeded in preserving that frozen man seen by Heuvelmans in 1968. Inside the freezer there was artificial ice and the man disappeared; he had a white skin and black hair. That man had a broken arm and his eyes had been shot. His upturned nosed showed two small cylindrical holes through which blew the air of the forest. Not everything is destroyed. Memories come back slowly. We did for some time know our predecessor. Let's not despair from seeing him again. Otherwise, we will seek him on planets where mankind has merely reached the stage of our precursor. Alternately, we will take more seriously page 24 of the 10th edition of Linnaeus's book, stupidly altered in 1788 in its 13th edition.

Author's Note

Shortly after having written the above, I learned of the sudden death of René Laurenceau, with whom I had spoken on the phone only a few days earlier (September 2010). Francophone readers may wish to consult the magazine *Bipedia,* edited by Francois de Sarre, and available online at http://cerbi.ldi5.com/rubrique.php3?id_rubrique=12. Laurenceau's remarkable contributions on the wild man are found in issues 7, 12, 14, 15, 17, 20 and 21.

Bibliography

Books

André, Jacques. 1995. *Etre médecin à Rome*. Petite Bibliothèque Payot, Paris, France.
Aroles, Serge. 2007. *L'énigme des enfants-loups*. Publibook, Paris, France.
Barloy, Jean-Jacques. 2007. *Bernard Heuvelmans, Un Rebelle de la Science*. Editions de l'œil du Sphinx, Paris, France.
Baudrimont, Albert-Frédéric. 2010. *Le Yéti démystifié*. P.R.N.G. Editions, Monein, France.
Bayanov, Dmitri. 1996. *In the footsteps of the Russian Snowman*. Crypto-Logos, Moscow.
Boas, Franz. 1995. *A Wealth of Thought* (edited by Aldona Jonaitis). University of Washington Press, Seattle, WA and London, UK.
Buffetaut, Eric. 1998. *Histoire de la paléonthologie*. Presses Universitaires de France, Paris, France.
Byrne, Peter. 1976. *The Search for Bigfoot*. Pocket Books, New York, NY.
———. 2013. *The Monster Trilogy Guidebook*. Hancock House Publishers, Surrey, B.C., Canada/Blaine, WA, U.S.A.
Cachel, Susan. 2006. *Primate and Human Evolution*. Cambridge University Press, Cambridge, UK.
Clews Parsons, Elsie. 2008. *Pueblo Indian Religion*. Vol. 2. University of Nebraska Press, Lincoln, NE.
Clottes, Jean and David Lewis-Williams. 2007. *Les chamanes de la préhistoire*. Editions du Seuil, Paris, France.
Coleman, Loren. 1989. *Tom Slick and the Search for the Yeti*. Faber & Faber, London, UK.
Coudray, Philippe. 2009. *Guide des animaux cachés*. Editions du Mont, Cazouls-les-Béziers, France.

Cremo, Michael A. and Richard L. Thomson. 1993. *Forbidden Archeology*. Bhaktivedanta Book Publishing Inc., Los Angeles, USA; Sydney, Australia; Stockholm, Sweden; Bombay, India.

Davy, Marie-Madeleine. 1978. *Encyclopédie des Mystiques*. Tome 4, Seghers, Paris, France.

Debenat, Jean-Paul. 2007. *Sasquatch et le Mystère des Hommes Sauvages*. Editions Le Temps Présent, Agnières, France.

———. 2009. *Sasquatch/Bigfoot and the Mystery of the Wild Man* (revised and translated from the preceding). Hancock House Publishers, Surrey, BC, Canada.

———. 2011. *A la Poursuite du Yéti*. Editions Le Temps Présent, Agnières, France

Debenath, André. 2006. *Néandertaliens et Cro-Magnons*. Editions Le Croît Vif, Paris, France.

Desimpelare, Jean-Paul and Elizabeth Martens. 2009. *Tibet: au-delà de l'illusion*. Editions Aden, Bruxelles, Belgique.

Dong, Paul. 1996. *China's Major Mysteries*. China Books, San Francisco, CA.

Dorst, Jean. 1995. *Les Animaux et le Sacré*. Albin Michel, Paris, France.

Durand-Tullou, Adrienne. 2002. *Un milieu de civilisation traditionnelle— le Causse de Blandes*. Editions du Beffroi, Millau, France.

Eberhart, George M. 2002. *Mysterious Creatures: a guide to cryptozoology*. Vol. 2. ABC-CLIO Inc., Santa Barbara, CA.

Eliade, Mircea. 1964. *Shamanism: archaic techniques of ecstasy*. Trans. W. Trask. Routledge and Kegan Paul, London, UK.

———. 1983. *Le chamanisme*, Payot, Paris, France.

——— (editor in chief). 1987. *Encyclopedia of religion*. Macmillan, New York, NY.

——— and Ioan P. Couliano. 1990. *Dictionnaire des religions*. Plon, Paris, France.

Frossard, Véra. 2004. *La Mémoire du Yéti*. L'Harmattan, Paris, France.

Gaup, Ailo. 1998. *Le tambour du chamane*. Editions du Reflet, Trouville-sur-Mer, France.

Gould, Stephen J. 1977. *Ever Since Darwin: reflections in natural history*. W.W. Norton, New York, NY.

———. 1997. *Darwin et les grandes énigmes de la vie*. Editions du Seuil, Paris, France.

Hart, Mickey. 1990. *Drumming at the Edge of Magic: A Journey into the Spirit of Percussion*. Lost Books, Little Elm, Texas.

Heuvelmans, Bernard. 1955. *Sur la piste des bêtes ignorées*. Plon, Paris, France.

———. 1995. *On the track of unknown animals* (translated and revised version of the above). Kegan Paul International, London, UK.

———— and Boris Porchnev. 1974. *L'Homme de Néanderthal est toujours vivant.* Plon, Paris, France.

de Heusch, Luc. 1995. "Possédés somnambuliques, chamans et hallucinés" in *La Transe et l'Hypnose* (ouvrage collectif), p.42, Imago, Paris, France.

Innes, Hammond. 1971. *Levkas Man.* Alfred A. Knopf, New York, NY.

Kalweit, Holger. 1992. *Shamans, Healers and Medicine Men.* Shambala, London, UK and Boston, MA.

Kerr, Philip. 1996. *Esau.* Chatto and Windus, London, UK.

Lall, Kesar. 1988. *Tales of the Yeti.* Pilgrims Book House, Thamel, Kathmandu, Népal.

Lall, Kesar. 1988. *Lore and Legend of the Yeti.* Pilgrims Book House, Thamel, Kathmandu, Népal.

Laurenceau. René. 2009. *Pâté d'alouette et de cheval.* Editions Amalthée, Nantes, France.

Lecomte-Tilouine, Marie. 1996. *Célébrer le pouvoir*, CNRS Editions, Paris, France.

LeNoël, Christian. 2005. *La Race Oubliée.* Tome 2. Chez l'auteur, 92, rue H. Lacroix, 83000 Toulon, France.

Le Quellec, Jean-Loïc. 1996. *Petit Dictionnaire de zoologie mythique.* Editions Entente, Paris, France.

L'Homme, Erik. 2010. *Des Pas dans la Neige.* Gallimard, Paris, France.

McCrone, John. 1992. *The Ape that Spoke.* Avon Books, New York, NY.

Mazdelstam Balzer, Marjorie, Ed. 1990. *Shamanism: Soviet studies of traditional religion in Siberia and Central Asia.* M. E. Sharpe Inc., Armonk, New York, NY.

Manoury, Pierre. 2007. *Encyclopédie du chamanisme.* Editions Trajectoire, Paris.

Markotic, Vladimir and Grover Krantz, Editors, 1984. *The Sasquatch and other Unknown Hominoids*, Western Publishers, Calgary, Canada.

Matthiessen Peter. 1978. *The Snow Leopard.* Viking Press, New York, NY.

Messner, Reinhold, 2000. *Yéti. Du mythe à la réalité.* Glénat, Grenoble, France.

————. 2000. *My Quest for the Yeti.* Saint Martin's Press, New York, NY.

Miller, Carey. 1987. *A Dictionary of Monsters and Mysterious Beasts.* Pan Books, London, UK.

Moskowitz Strain, Kathy. 2008. *Giants, Cannibals and Monsters.* Hancock House Publishers, Surrey, BC, Canada.

Nabokov, Peter. 2008. *Là où frappa la Foudre.* Editions Albin Michel, Paris.

Nicholson, S. 1987. *Shamanism: an Expanded View of Reality.* Theosophical Publishing House, Wheaton, Illinois.

Pegg, Carole. 2001. *Mongolian Music, Dance and Oral Narrative.* University of Washington Press, Seattle and London.
Perrin, Michel. 2007. *Voir les yeux fermés.* Editions du Seuil, Paris.
Picq, Pascal PICQ. 2007. *Nouvelle Histoire de l'Homme.* Editions Perrin, Paris.
Raibaud, Marie-Paule. 2004. *Femmes d'une autre Chine.* Editions du Mont, Cazouls-les-Béziers, France.
Rawicz, Slavomir. 2002. *A Marche Forcée (A pied du Cercle Polaire à l'Himalaya—1941–1942).* Editions Phébus, Paris.
———. 1997. *The Long Walk.* The Lyons Press, Guilford, CT, USA.
Riggs, Rob. 2001. *In the Big Thicket (on the Trail of the Wild Man).* Paraview Press, New York, NY.
Robert, Paul. 2001. *Le Grand Robert de la Langue Française.* 6 volumes. Dictionnaires Le Robert, Paris.
Rossi, Lorenzo. 2008. *Gli Ultimi Neandertal.* Editions Boopen, Italy.
Rouaud, Jean. 2007. *Préhistoires.* Editions Gallimard, Paris.
Roussot, Alain. 1997. *L'Art Préhistorique.* Editions Sud-Ouest, Bordeaux.
de Sales, Anne. 1991. *Je suis né de vos jeux de tambours (La religion chamanique des Magars du Nord).* Société d'Ethnologie, Nanterre, France.
Saunders, Nicholas J. 1995. *Animal Spirits.* Little, Brown, Boston, MA.
Saunders, Nicholas J. 1995. *Les Animaux et le sacré.* Editions Albin Michel, Paris.
Servier, Jean. 1980. *L'homme et l'invisible.* Editions Imago, Paris.
Shiel, Lisa A. 2006. *Backyard Bigfoot.* Slipdown Mountain Publications LLC, Lake Linden, Michigan.
Silverberg, Robert. 1984. *Gilgamesh the King.* Arbor House.
Stein, Gordon. 1993. *Encyclopedia of Hoaxes.* Gale Research Inc., Detroit, Washington DC, London.
Tchernine, Odette, Ed. 1958. *Explorers' and Travellers' Tales.* Jarrolds Publishers, London, UK.
———. 1971. *In Pursuit of the Abominable Snowman.* Taplinger Publishing Co., New York, NY.
——— and Gerald Moore. 1976. *The Singing Dust.* Neville Spearman, London, UK.
Van Grasdorff, Gilles. 2004. *L'Attrapeur de pluie.* Editions Jean-Claude Lattès, Paris.
———. 2006. *La nouvelle histoire du Tibet.* Editions Perrin, Paris.
Xiao, Xiaoming (chief editor). 2007. *Les ethnies minoritaires.* Editions en Langues étrangères, Beijing, Chine.
Zhang, Qing. 2005. 野人. 2005. *Yeren.* Yuan Fang Publishing House, Hohhot, Inner Mongolia, China. (Popular Chinese language work on the Yeren).

Articles (magazines, collaborations, websites)

Arnaud, Bernadette. 2009. "A chacun son Yéti," pp. 78 à 82; "Des siècles de recherches, et toujours pas de preuves," pp. 83 à 85, in *Sciences et Avenir*, Paris, juillet.

Coleman, Loren E. et Mark A. Hall. 1975. "L'Abominable Homme des Etats-Unis," pp. 145–166, in *Le Livre de l'Inexplicable* de Jacques Bergier et le Groupe Info, France-Loisirs, Paris.

Forth, Gregory. 2007. "Images of the Wildman Inside and Outside Europe," pp. 261–281, in *Folklore*, vol. 118, 3 Decembre 2007.

Grison, Benoît. 2003. "Psychologie Soviétique, Histoire sociale et Anthropogenèse: la « lutte pour les troglodytes » de Boris Porchnev," in *Institut Virtuel de Cryptozoologie*: http://pagesperso-orange.fr/cryptozoo/, juin.

de Heush, Luc. 1995. "Possédés somnambuliques, chamans et hallucinés," in *La Transe et l'Hypnose*, ouvrage collectif sous la direction de Didier Michaux, Editions Imago, Paris.

Heuvelmans, Bernard. 1996. "Le chimpanzé descend-il de l'homme?," pp. 86–91, in *Planète* n°31, nov./déc.

Koffmann, Marie-Jeanne. 1991. "L'Almasty, Yéti du Caucase", Archeologia, n°269, juin.

———. 1992. "L'Almasty du Caucase - Mode de vie d'un humanoïde", Archeologia, n°276, février.

———. 1994. "Les Hominoïdes Reliques dans l'Antiquité" (1ère partie), Archeologia, n°307, décembre.

———. 1995. "Les Hominoïdes Reliques dans l'Antiquité" (2ème partie), Archeologia, n°308, janvier.

Laurenceau, René. 2003. "Nocturnisme et Bestialité," in *Bipédia* (revue du Centre d'Etudes et de Recherches sur la Bipédie Initiale: C.E.R.B.I.), vol. 15.1, article posted 30 June on http://cerbi.ldi5.com/

Raynal, Michel. 1993. "Les Néanderthaliens Reliques, des Pyrénées au Pakistan," in *Bipédia*, vol. 10, juin: http://cerbi.ldi5.com/

White, Scott. 2002. "Les Empreintes dans la Forêt," in *Bipédia*, vol. 20, Jan: http://cerbi.ldi5.com/

Photo Credits

All from the author's collection, except black and white photos (pages 21, 22 and 34) from *Paris-Match* No. 475, 17 May 1958, reproduced with permission of *Paris-Match*.

Index

The various names for wild men are in *italics*;
bold numbers indicate photographs.

A
Ainu, 80–81
Allsop, Mike, 28
almas, almasty, 6, 82, 88–90, 95 100, 106–110, 112
   diet 120–121
   as domestic servant, 127
Altai range, 78
Amala and Kamala. *See* wolf girls of Midnapore
Anasazi, 150, 153,
bigfoot, **153**
Arun River, 21, 25
Australopithecus, 58, 61, 139

B
Bal des Ardents, 111–2
Ban/van manas, 16
Bannon, Brandon, 102
Baradyine, Professor, 88
Barun Hüre monastery, 95
Baruun-Urt monastery, 90
Bayanov, Dmitri, **108**, 113, 140
Beauman, E.B., 27
Behn, Mira, 16
Bennett, Ralph, 36
Bhutan, 17, 66, 98
Big Footprints: A Scientific Enquiry into the Reality of Sasquatch, 130
Bipedia, 74, 158, 159, 161
Bon religion, 67, 80, 94
Bourtsev, Igor, **101, 102,** 113
braiding manes, 111–13
Britton, Sydney, 29
Buriats, 39, 79
Bury, C.K. Howard, 14
Bykova, Maya, 113
Byrne, Peter, 13–15 20, **22**
Byrne, Brian, 20–25, **21, 22**

C
Cachel, Suzanne, 134
Campbell, Joseph, 37–40
Cato the Elder, 37
Caucasus, 106–**109**, 120-4, 127, 140, 141, 145, 159
Chalfant Valley Group, 154
Chemo, 64–5, 70, 72, 73, 95
Chinese langur, 136, **139**
Chiwong Gompa, 67
chuchunyas, 78–80, 158–9
Chung-Min Chen, 137
Coudray, Philippe, 17, 152–3

D
Daily Mail expedition, 20, 27, 32
damaru. See skull drum
dancing sorcerer, 38–9
Dart, Raymond, 60, 61
d'Avergne, Captain, 23
David, Armand, 136, 142
David-Néel, Alexandra, 34, 50–1, 91–2

Dawson, Charles, 60, 61
de Milleville, René, 56
de Sales, Anne, 41
de Sarre, François, 74–5, 139–40
Debenet, Gilles, 148
Diamond, Jared, 61
Dinopithecus nivalis, 31, 33
divine lake. *See* Yulung Lhantso
Domovoi, 158
Dong, Paul, 131
Dumje festival, 53
Dyhrenfurth, Norman, 20, **22**
Dzonokwa, 152, **154**

E

Eichinger, Franz, 93
Eliade, Mircea, 45
Enkidu, 115, 116–7, 119
Epic of Gilgamesh. *See* Gilgamesh
Esau, 51, 57–61, 63, 119, 160
ethnomedecine, 37
Everest Reconnaissance Expedition, 14, 16

F

feral children, 121–3
Festival of the Bear, 54
Froissart, Jean, 112
Frossard, Vera, 49–51, 54–5
Fusch, Ed, 145

G

Gangtey Gompa, 69, 95
giant panda, 137, 142
Gigantopithecus, 57, 58, 139
Gilgamesh 115–7, 119
Gobi Desert, 89
Gody, 44–5
golden monkey, 136–7
Goodall, Jane, 58
gorillaï, 115
Grotte des Trois Frères, 38
Guangxi (province), 130, 131
Gyalzen, 55, 56

H

Hanno, 115
Hart, Mickey, 39, 40
Hergé, 29–32, 54
Heath-Stubbs, John, 105
Hero With a Thousand Faces, 37
Heuvelmans, Bernard, **25**, 26, 27, 29, 66, 140, 148
Hillary, Edmund 15, **25**, 26, 63
Hindu Kush mountains, 84
Holton, George, 20
Homo
  erectus, 58, 134, 140, **141**
  ergaster, 58
  habilis, 58
  pongoides, 141
  robustus, 57–58
  rudolfensis, 58
  sapiens, 58, 114, 117, 141, 144, 145
  sylvestris, 114
  troglodytes, 114
  vertex, 60
Hooker, John Dalton, 19
Hooker, Joseph, 17
Hopi, 150–7
Hu Hongzing, 137
Hunt, John, 15
Hutchison, Robert, 55

I

iboga, 41
In the Footsteps of the Russian Snowman, 108, **113**
Indian Mountaineering School, 15
Innes, Hammond, 145
Izzard, Ralph, 27

J

jamara, 47
Johnson, Kirk, 20

K

Kabardino–Balkaria, 107, 124
kachinas, 151–2, 154–6
kalash, 47
Karakantz, Georgios, 148

Karapetian, Vazghen, 108–110
Karmapa, 149–51, 156
Kearney, Joyce, 152, 154–6
Kerr, Philip, 51, 57, 58, 60, 63
Khakhlov, V.A., 82–4
khoun-gouressou, 83
Khwit, **102**
Koffmann, M.J., 24, **101**, 105ff, 108, 111, 114ff
Krantz, Grover, **104**, 130–3
Krief, Sabrina, 36
ksy-gyik, 82
Kumbila, 53
!Kung Bushmen, 37

L

Lake Baikal, 78
Lake Zaisan, 82
Lall, Kesar, 93
Lascaux, 114, **115**
L'Attrapeur de Pluie, 149
Laurenceau, René, 158
Leakey, Louis, 58, 61, 145
Leroi-Gourhan, 57
Lhokpa Pass, 14
Lucretius, 37
lung-gom-pa lamas, 92

M

Machkovtsev, A.A., **108**
Magars, 41
Mani Rimdu, 67
maps
    Caucasus area, 109
    China and surrounding countries, 71
    Himalayan realm, 18
    Major Wild Man Sightings in China, 135
    Mongolia, 87
    Nepal, 14
Matthiessen, Peter, 32
Menlung Glacier, 15, 16
Messner, Reinhold, 62–5, 67–75, 78, 81, 87, 90, 95–6
    Günther (brother), 63
Metah-Kangmi, 22

Miao people, 130–3
Milarepa, 34
mi-gö, mighu, 66
mi-gou, 31
mi-the, 32
Mongolian Academy of Sciences, 88
moumieu, 85
My Quest for the Yeti, 63

N

Neanderthal, 24, 26, 57, 58, 78, 86, 117, 140, 145, 146, 148, 159, 160, 161
Nepal, **14**, 20–7, 31, 41, 47, 51, 56, 66
Norgay, Tenzing, 15
Nutall, George, 61

O

Ogre of the Mountains, 134
Okladnikov, A.P., 78
Obruchev, S.V., 24
Ogden, Rob, 28
Osman Hill, W.C., 26

P

Padmasambara, 149–151
Paï Hsin, 18, 19
Pallix, Silvain, 124, 125
Pamir Mountains, 18
Panchenko, Gregory, 111, 112–3
Pangboche
    hand, 25–6, 28
    monastery (Gompa), 25, 32, 33, 93
    scalp, 25
Paranthropus
    Boisei, 58
    robustus, 57–58, 140
Paris Match, 20
Patterson and Gimlin, 125
Peking Daily, 17
Pichon, Jean-Charles, 45
Piltdown Man, 61
Piveteau, Prof., 120
Plawinski, W., **89**
Pliny the Elder, 37
Poirier, Frank, 137–8

Porchnev, Boris, 78, 83–5, 88–9, 105, **108**, 111, 113
potlatch, 53
Pongo erectus, 133
Primihomo asiaticus, 84
Prince Peter of Greece, 15
Prjevalsky, N., 83
Prospector Ed. *See* Fusch, Ed

R

Rain Catcher, The. See L'Attrapeur de Pluie
rakshas, 13
Rawicz, Slavomir, 30, 35
Rhinopithecus roxellanae. See Chinese langur
Richen, Prof., 88–9
Royal Geographical Society, 14
Royal Zoological Society of Scotland, 28
russalka, 158–9
Russell, Gerald, **22**, 23, 27, 66

S

sangbai-dagpo, 32
Sanglakh Range, 84
Saturnalia, 43, 46
Schaller, George, 64
Seminar for Research on Relic Hominids, 109
Seneca, 36–37
serow goat, 26, 33, 99
Shackley, Myra, 28, 136, 143, 158
Shakespeare, 43–4, 111, 113
shaman/shamanism, 36, 38, 39, 40, 41, 42, 53, 56, 67, 68, 79, 80, 85, 88, 95, 96, 127, 131, 146, 156, 159
Shennong, 142
Shennongjia Forest, 134–43, 148
Shipton, Eric, 15, 16, 64
shookp/sogpa, 13
Silverberg, Robert, 116
Sikkim, 14–15
skull drum, 40
Slick, Tom, 15, 20, 24, 26, 28
Smoline, P.P., **108**

snow monkey, **137**. *See also* golden monkey
Sosar Gompa, 73, 95
Stewart, James, 25
Stonor, Charles, 32–33
Sumerians, 116

T

tahi, 88
Tang Dynasty, 134
Tarbagatai Mountains, 82
Tarchen, 67
Tarthang Tulku, 40, 42
Tchernine, Odette, 16, 31, 88, 95, **106**
thod raga. See skull drum
Tien Shan, 31
Tintin, 29, 31, 32, 54, 144
Tombazi, N.A., 14
toumo, 51, 92
Track Record, The, 153, 154
Tranier, Michel, 56

U

Van Grasdorff, Gilles, 148
Verkhoyansk Mountains, 78
Victor of Aveyron, 122

W

Wang Zelin, 134
Ward, Michael, 16
wolf girls of Midnapore, 121
Wu Dingliang, 138

Y

yak, 71, 76
yakutia, 79
yati, 93
yeren, **103, 104**, 128, 130–2, 138–9
Yeren sinensis, 133
Yerpa Valley, 93–4
yeti
    as a god, 52
    diet, 26, 32, 66, 70
    feet, 30, 99
    hand, 26–26, **99**
    hermit, 91

kidnapper, 55
killing yaks, 50, 95
types, 26, 140
scalps, **25,** 26, 33, **99**
pelt, 69, 72
Yeti Reserve. *See* Shennongjia Forest
Yuan Mei, 134
Yuan Zhenxin, 141–2
Yule, 43, 45
Yulung Lhantso, 70

Z

Zana, **102**
Zdorik, B.M., 84
Zemu Glacier, 14
Zhou Guoxing, 104, 131, **142**
Zuni, 153, 155

About the Author

Jean-Paul Edouard Debenat was born in 1943 in the "département" of Vendée in the west of France. His father, a railway stationmaster, had to move to a different town every four to five years, from the La Rochelle area to Brittany. Jean-Paul developed a taste for traveling, probably initiated by his (semi-clandestine) trips on board freight trains, and he dreamed of becoming an airline pilot so he could travel the world. His dream was shattered when a medical exam revealed weak eyesight, so Jean-Paul turned to the study of English at the University of Rennes. This would still allow him to visit many countries. He taught French in Glasgow, Scotland and at the University of Notre-Dame, South Bend, Indiana. He wrote his dissertation on science fiction writer Robert A. Heinlein for a PhD in Comparative Literature (La Sorbonne, Paris, 1984); and he studied the history of religions at the renowned Ecole Pratique des Hautes Etudes in Paris.

Jean-Paul enjoyed a long-lasting friendship with novelist and mythologist Jean-Charles Pichon; and he became closely acquainted with Dr. Bernard Heuvelmans, whom he used to visit at his home in Le Bugue, near Lascaux, and then at Bernard's last residence in Le Vésinet, near Paris. Pichon and Heuvelmans knew each other and their erudition stimulated Jean-Paul's creativity, leading him to spend a sabbatical in the state of Washington (1994) studying the bigfoot/sasquatch phenomenon. He was able to meet *in situ* most of the major North American researchers at the time. His first book on the topic was published in French (Editions JMG – Le Temps Présent, 2007), before being translated by Dr. Paul LeBlond under the title

Jean-Paul Debenat is holding a rattle made of deer hooves in Quebec, Canada. In front of him, on the table are two drums and two small Native American rattles. By his right hand are a prayer wheel and a Tibetan singing bowl.

Sasquatch/Bigfoot and the Mystery of the Wild Man (Hancock House Publishers, Canada/USA, 2009).

After several trips to the province of Sichuan, China, where he visited his son Julien and daughter-in-law Mao Xi, Jean-Paul decided to investigate the findings of the researchers who devoted their energy tracking the mysterious "wild men of Asia," which is the title of his second book. It was published in French as *A la Poursuite du Yéti* [In Pursuit of the Yeti] (Editions JMG – Le Temps Présent, 2009). Jean-Paul Debenat has lectured on cryptozoology as well as on aviation, in France, Belgium, Italy, Great Britain, USA and China.

Presently, Jean-Paul is completing a book on the history of aviation—his second book in the domain—to be published by Editions Bleu-Ciel under the title, *Avions et Pilotes d'Exception* [Outstanding Planes and Pilots].

About the Translator

Paul LeBlond is a Canadian ocean scientist, born in Quebec City in 1938 and now resides on Galiano Island, British Columbia. A graduate of McGill University and the University of British Columbia, LeBlond taught physics and oceanography at the University of British Columbia where he is now an emeritus professor. His work on ocean waves and currents has taken him to research institutes in Germany, France and Russia and was applied to practical problems through his industrial consulting activities and membership in fisheries conservation councils.

Parallel to his academic work, he developed a keen interest in unidentified marine animals, inspired by the work of Bernard Heuvelmans. He participated in the creation of the International Society of Cryptozoology in 1982 and was a co-founder the British Columbia Scientific Cryptozoology Club in 1989. LeBlond is a Fellow of the Royal Society of Canada and of the Canadian Meteorological and Oceanographic Society, as well as a foreign member of the Academy of Natural Sciences of the Russian Federation.

The author (left) with friend, and future translator, Dr. Paul LeBlond, UBC Vancouver, July 7, 1995.
PHOTO: Marie-Agnès Debenat

Also by **Jean-Paul Debenat**

Sasquatch/Bigfoot
and the Mystery of the Wild Man
Cryptozoology and Mythology
in the Pacific Northwest
• • •
ISBN 978-0-88839-685-3
5½ x 8½ inches, sc, 428 pages
132 photos, 32-page color photo section

www.hancockhouse.com | sales@hancockhouse.com

Other HANCOCK HOUSE cryptozoology titles

Best of Sasquatch Bigfoot
John Green
0-88839-546-9
8½ x 11, sc, 144 pp

Bigfoot Encounters in Ohio
C. Murphy, J. Cook, G. Clappison
0-88839-607-4
5½ x 8½, sc, 152 pp

Bigfoot Encounters in New York & New England
Robert Bartholomew
Paul Bartholomew
978-0-88839-652-5
5½ x 8½, sc, 176 pp

Bigfoot Film Controversy
Roger Paterson,
Christopher Murphy
0-88839-581-7
5½ x 8½, sc, 240 pp

Bigfoot Film Journal
Christopher Murphy
0-88839-658-7
8½ x 11, sc, 106 pp

Bigfoot Research: The Russian Vision
Dmitri Bayanov
978-0-88839-706-5
5½ x 8½, sc, 432 pp

Bigfoot Sasquatch Evidence
Dr. Grover S. Krantz
0-88839-447-0
5½ x 8½, sc, 348 pp

Giants, Cannibals & Monsters
Kathy Moskowitz Strain
0-88839-650-3
8½ x 11, sc, 288 pp

Hoopa Project
David Paulides
0-88839-653-2
5½ x 8½, sc, 336 pp

In Search of Giants
Thomas Steenburg
0-88839-446-2
5½ x 8½, sc, 256 pp

The Locals
Thom Powell
0-88839-552-3
5½ x 8½, sc, 272 pp

Monster Trilogy Guidebook
Peter Byrne
978-0-88839-723-2
5½ x 8½, sc, 176 pp

Know the Sasquatch
Christopher Murphy
978-0-88839-657-0
8½ x 11, sc, 320 pp

Raincoast Sasquatch
J. Robert Alley
978-0-88839-508-5
5½ x 8½, sc, 360 pp

Sasquatch: The Apes Among Us
John Green
0-88839-123-4
5½ x 8½, sc, 492 pp

Sasquatch in British Columbia
Christopher Murphy
0-88839-718-8
5½ x 8½, sc, 528 pp

Tribal Bigfoot
David Paulides
978-0-88839-687-7
5½ x 8½, sc, 336 pp

Who's Watching You?
Linda Coil Suchy
978-0-88839-664.8
5½ x 8½, sc, 408 pp

Yale & the Strange Story of Jacko the Ape-boy
Christopher Murphy
978-0-88839-712-6
5½ x 8½, sc, 48 pp

www.hancockhouse.com | sales@hancockhouse.com